The NASA STI Program Office ... in Profile

Since its founding, NASA has been dedicated to the advancement of aeronautics and space science. The NASA Scientific and Technical Information (STI) Program Office plays a key part in helping NASA maintain this important role.

The NASA STI Program Office is operated by Langley Research Center, the lead center for NASA's scientific and technical information. The NASA STI Program Office provides access to the NASA STI Database, the largest collection of aeronautical and space science STI in the world. The Program Office is also NASA's institutional mechanism for disseminating the results of its research and development activities. These results are published by NASA in the NASA STI Report Series, which includes the following report types:

- TECHNICAL PUBLICATION. Reports of completed research or a major significant phase of research that present the results of NASA programs and include extensive data or theoretical analysis. Includes compilations of significant scientific and technical data and information deemed to be of continuing reference value. NASA's counterpart of peer-reviewed formal professional papers but has less stringent limitations on manuscript length and extent of graphic presentations.

- TECHNICAL MEMORANDUM. Scientific and technical findings that are preliminary or of specialized interest, e.g., quick release reports, working papers, and bibliographies that contain minimal annotation. Does not contain extensive analysis.

- CONTRACTOR REPORT. Scientific and technical findings by NASA-sponsored contractors and grantees.

- CONFERENCE PUBLICATION. Collected papers from scientific and technical conferences, symposia, seminars, or other meetings sponsored or cosponsored by NASA.

- SPECIAL PUBLICATION. Scientific, technical, or historical information from NASA programs, projects, and mission, often concerned with subjects having substantial public interest.

- TECHNICAL TRANSLATION. English-language translations of foreign scientific and technical material pertinent to NASA's mission.

Specialized services that complement the STI Program Office's diverse offerings include creating custom thesauri, building customized databases, organizing and publishing research results ... even providing videos.

For more information about the NASA STI Program Office, see the following:

- Access the NASA STI Program Home Page at http://www.sti.nasa.gov/STI-homepage.html

- E-mail your question via the Internet to help@sti.nasa.gov

- Fax your question to the NASA Access Help Desk at (301) 621-0134

- Telephone the NASA Access Help Desk at (301) 621-0390

- Write to:
  NASA Access Help Desk
  NASA Center for AeroSpace Information
  7121 Standard Drive
  Hanover, MD 21076-1320

NASA/TP—2001–209990

Total Solar Eclipse of 2002 December 04

Fred Espenak
Goddard Space Flight Center, Greenbelt, Maryland

Jay Anderson
Environment Canada, Winnipeg, Manitoba

National Aeronautics and
Space Administration

Goddard Space Flight Center
Greenbelt, Maryland 20771

September 2001

Available from:

NASA Center for AeroSpace Information 7121 Standard Drive Hanover, MD 21076-1320 Price Code: A17	National Technical Information Service 5285 Port Royal Road Springfield, VA 22161 Price Code: A10

Preface

This work is the latest in a series of NASA publications containing detailed predictions, maps and meteorological data for future central solar eclipses of interest. Published as part of NASA's Technical Publication (TP) series, the eclipse bulletins are prepared in cooperation with the Working Group on Eclipses of the International Astronomical Union and are provided as a public service to both the professional and lay communities, including educators and the media. In order to allow a reasonable lead time for planning purposes, eclipse bulletins are published 18 to 24 months before each event.

Single copies of the bulletins are available at no cost by sending a 9 x 12-inch self-addressed stamped envelope with postage for 12 oz. (340 g.). Detailed instructions and an order form can be found at the back of this publication.

The 2002 bulletin uses the World Data Bank II (WDBII) mapping data base for the path figures. WDBII outline files were digitized from navigational charts to a scale of approximately 1:3,000,000. The data base is available through the Global Relief Data CD-ROM from the National Geophysical Data Center.

The highest detail eclipse maps are constructed from the Digital Chart of the World (DCW), a digital database of the world developed by the U.S. Defense Mapping Agency (DMA). The primary sources of information for the geographic database are the Operational Navigation Charts (ONC) and the Jet Navigation Charts (JNC). The eclipse path and DCW maps are plotted at a scale of 1:2,000,000 in order to show roads, cities and villages, lakes and rivers, suitable for eclipse expedition planning.

The geographic coordinates data base includes over 90,000 cities and locations. This permits the identification of many more cities within the umbral path and their subsequent inclusion in the local circumstances tables. These same coordinates are plotted in the path figures and are labeled when the scale allows. The source of these coordinates is Rand McNally's The New International Atlas. A subset of these coordinates is available in a digital form which we've augmented with population data.

The bulletins have undergone a great deal of change since their inception in 1993. The expansion of the mapping and geographic coordinates data bases have significantly improved the coverage and level of detail demanded by eclipse planning. Some of these changes are the direct result of suggestions from our readers. We strongly encourage you to share your comments, suggestions and criticisms on how to improve the content and layout in subsequent editions. Although every effort is made to ensure that the bulletins are as accurate as possible, an error occasionally slips by. We would appreciate your assistance in reporting all errors, regardless of their magnitude.

We thank Dr. B. Ralph Chou for a comprehensive discussion on solar eclipse eye safety. Dr. Chou is Professor of Optometry at the University of Waterloo and he has over twenty-five years of eclipse observing experience. As a leading authority on the subject, Dr. Chou's contribution should help dispel much of the fear and misinformation about safe eclipse viewing.

Dr. Joe Gurman (GSFC/Solar Physics Branch) has made this and previous eclipse bulletins available over the Internet. They can be read or downloaded via the World Wide Web from Goddard's Solar Data Analysis Center eclipse information page: *http://umbra.nascom.nasa.gov/eclipse/*.

In 1996, Fred Espenak developed the *NASA Eclipse Home Page*, a web site which provides general information on every solar and lunar eclipse occurring during the period 1951 through 2050. An online catalog also lists that date and characteristics of every solar and lunar eclipse from 2000 BC through AD 3000. The URL for the site is: *http://sunearth.gsfc.nasa.gov/eclipse/eclipse.html*.

In addition to the general information web site above, a special web site devoted to the 2002 total solar eclipse has been set up: *http://sunearth.gsfc.nasa.gov/eclipse/TSE2002/TSE2002html*. It includes supplemental predictions, figures and maps which could not be included in the present publication.

Since the eclipse bulletins are of a limited and finite size, they cannot include everything needed by every scientific investigation. Some investigators may require exact contact times which include lunar limb effects or for a specific observing site not listed in the bulletin. Other investigations may need customized predictions for an aerial rendezvous or from the path limits for grazing eclipse experiments. We would like to assist such investigations by offering to calculate additional predictions for any professionals or large groups of amateurs. Please contact Espenak with complete details and eclipse prediction requirements.

We would like to acknowledge the valued contributions of a number of individuals who were essential to the success of this publication. The format and content of the NASA eclipse bulletins has drawn heavily upon over 40 years of eclipse *Circulars* published by the U. S. Naval Observatory. We owe a debt of gratitude to past and present staff of that institution who have performed this service for so many years. The many publications and algorithms of Dr. Jean Meeus have served to inspire a life-long interest in eclipse prediction. We thank Francis Reddy, who helped develop the original geographic data base and to Rique Pottenger for his assistance in expanding the data base to over 90,000 cities. Peter Anderson of the Astronomical Association of Queensland provided a great deal of information about Australia. Nick Zambatis and Peter Tiedt contributed information about cloud and weather in Kruger National Park. GPS coordinates of key locations throughout Kruger National Park were generously provided by Peter Tiedt. Prof. Jay M. Pasachoff reviewed the manuscript and offered many helpful suggestions. Internet availability of the eclipse bulletins is due to the efforts of Dr. Joseph B. Gurman. The support of Environment Canada is acknowledged in the acquisition of the weather data.

Permission is freely granted to reproduce any portion of this publication, including data, figures, maps, tables and text. All uses and/or publication of this material should be accompanied by an appropriate acknowledgment (e.g. - "Reprinted from *Total Solar Eclipse of 2002 December 04* Espenak and Anderson, 2001"). We would appreciate receiving a copy of any publications where this material appears.

The names and spellings of countries, cities and other geopolitical regions are not authoritative, nor do they imply any official recognition in status. Corrections to names, geographic coordinates and elevations are actively solicited in order to update the data base for future eclipses. All calculations, diagrams and opinions are those of the authors and they assume full responsibility for their accuracy.

	Fred Espenak NASA/Goddard Space Flight Center Planetary Systems Branch, Code 693 Greenbelt, MD 20771 USA		Jay Anderson Environment Canada 123 Main Street, Suite 150 Winnipeg, MB, CANADA R3C 4W2
email: FAX:	espenak@gsfc.nasa.gov (301) 286-0212	email: FAX:	jander@cc.umanitoba.ca (204) 983-0109

Current and Future NASA Solar Eclipse Bulletins

NASA Eclipse Bulletin	RP #	Publication Date
Annular Solar Eclipse of 1994 May 10	*1301*	*April 1993*
Total Solar Eclipse of 1994 November 3	*1318*	*October 1993*
Total Solar Eclipse of 1995 October 24	*1344*	*July 1994*
Total Solar Eclipse of 1997 March 9	*1369*	*July 1995*
Total Solar Eclipse of 1998 February 26	*1383*	*April 1996*
Total Solar Eclipse of 1999 August 11	*1398*	*March 1997*
NASA Eclipse Bulletin	TP #	Publication Date
Total Solar Eclipse of 2001 June 21	*1999-209484*	*November 1999*
Total Solar Eclipse of 2002 December 04	*2001-209990*	*September 2001*
- - - - - - - - - - - future - - - - - - - - - - -		
Annular and Total Solar Eclipses of 2003	—	*2002*
Annular and Total Solar Eclipses of 2005	—	*2003*
Total Solar Eclipse of 2006 March 29	—	*2004*

Table of Contents

Eclipse Predictions 1
    Introduction 1
    Umbral Path And Visibility 1
Maps of the Eclipse Path 2
    Orthographic Projection Map of the Eclipse Path 2
    Equidistant Conic Projection Map of the Eclipse Path 2
    Detailed Maps of the Umbral Path 3
Elements, Shadow Contacts and Eclipse Path Tables 3
Local Circumstances Tables 5
Estimating Times of Second and Third Contacts 6
Mean Lunar Radius 7
Lunar Limb Profile 7
Limb Corrections to the Path Limits: Graze Zones 9
Saros History 10
Weather Prospects for the Eclipse 12
    Introduction 12
African Weather Prospects 12
    African Overview 12
    The Countryside 12
    Large Scale Weather Patterns 13
    Cloud, Temperature, Rainfall, Winds 15–16
    Africa Summary 16
Australian Weather Prospects 16
    Australia Overview 16
    The Countryside 16
    Large Scale Weather Patterns 17
    Cloud, Temperature 18
    Rainfall, Winds 19
    Australia Summary 19
    Eclipse Viewing on the Water 19
    Weather Web Sites 20
Observing the Eclipse 21
    Eye Safety and Solar Eclipses 21
    Sources for Solar Filters 23
    IAU Solar Eclipse Education Committee 23
    Eclipse Photography 24
    Sky at Totality 26
    Contact Timings from the Path Limits 27
    Plotting the Path on Maps 27
IAU Working Group on Eclipses 28
Eclipse Data on Internet 28
    NASA Eclipse Bulletins on Internet 28
    Future Eclipse Paths on Internet 29
    Special Web Site for 2002 Solar Eclipse 29
Predictions for Eclipse Experiments 29
Algorithms, Ephemerides and Parameters 29
Bibliography 31
    References 31
    Meteorology 31
    Eye Safety 32
    Further Reading 32

LIST OF FIGURES AND TABLES

Figures

Figure 1: Orthographic Projection Map of the Eclipse Path 33
Figure 2: The Eclipse Path Through Africa 34
Figure 3: The Eclipse Path Through Western Africa 35
Figure 4: The Eclipse Path Through Eastern Africa 36
Figure 5: 2002 Eclipse Path – Western Angola 37
Figure 6: 2002 Eclipse Path – Central Angola 38
Figure 7: 2002 Eclipse Path – Eastern Angola 39
Figure 8: 2002 Eclipse Path – Caprivi Strip 40
Figure 9: 2002 Eclipse Path – Botswana & Zimbabwe 41
Figure 10: 2002 Eclipse Path – Zimbabwe & South Africa 42
Figure 11: 2002 Eclipse Path – Mozambique 43
Figure 12: The Eclipse Path Through Australia 44
Figure 13: The Eclipse Path Through Southern Australia 45
Figure 14: 2002 Eclipse Path – Ceduna, Australia 46
Figure 15: 2002 Eclipse Path – Woomera, Australia 47
Figure 16: 2002 Eclipse Path – Lyndhurst, Australia 48
Figure 17: 2002 Eclipse Path – Tickalara, Australia 49
Figure 18: The Lunar Limb Profile At 06:15 UT 50
Figure 19: Global Pressure and Weather Systems 51
Figure 20: Map of Mean Cloudiness 51
Figure 21: Graph of Mean Cloudiness Along the Eclipse Path 52
Figure 22: Clear Skies in Australia 52
Figure 23: Spectral Response of Some Commonly Available Solar Filters 53
Figure 24: The Sky During Totality as Seen From Center Line at 06:15 UT 54

Tables

Table 1: Elements of the Total Solar Eclipse of 2002 December 04 55
Table 2: Shadow Contacts and Circumstances 56
Table 3: Path of the Umbral Shadow 57
Table 4: Physical Ephemeris of the Umbral Shadow 58
Table 5: Local Circumstances on the Center Line 59
Table 6: Topocentric Data and Path Corrections – Lunar Limb Profile 60
Table 7: Mapping Coordinates for the Umbral Path – Africa 61
Table 8: Mapping Coordinates for the Umbral Path – Australia 62
Table 9: Mapping Coordinates for the Zones of Grazing Eclipse – Africa 63
Table 10: Mapping Coordinates for the Zones of Grazing Eclipse – Australia 64
Table 11: Local Circumstances for Africa: Angola – Benin 65
Table 12: Local Circumstances for Africa: Botswana – Equatorial Guinea 66
Table 13: Local Circumstances for Africa: Ethiopia – Mali 67
Table 14: Local Circumstances for Africa: Mayotte – Somalia 68
Table 15: Local Circumstances for Africa: South Africa 69
Table 16: Local Circumstances for Africa: Sudan – Zimbabwe 70
Table 17: Local Circumstances for Antarctica, Indonesia & Indian Ocean 71
Table 18: Local Circumstances for Australia & New Zealand 72
Table 19: Solar Eclipses of Saros Series 142 73
Table 20: African and Australian Weather Statistics 74
         Key to Table 20 75
Table 21: 35mm Field of View and Size of Sun's Image 76
Table 22: Solar Eclipse Exposure Guide 76

Request Form for NASA Solar Eclipse Bulletins 77

Eclipse Predictions

Introduction

On Wednesday, 2002 December 04, a total eclipse of the Sun will be visible from within a narrow corridor which traverses the Southern Hemisphere. The path of the Moon's umbral shadow begins in the South Atlantic, crosses southern Africa and the Indian Ocean, and ends at sunset in southern Australia. A partial eclipse will be seen within the much broader path of the Moon's penumbral shadow, which includes the southern two thirds of Africa, Antarctica, Indian Ocean and Australia (Figure 1).

Umbral Path And Visibility

The eclipse begins in the South Atlantic where the Moon's umbral shadow first touches down on Earth at 05:50 UT (Figure 2). Along the sunrise terminator, the duration is only 26 seconds as seen from the center of the 31 kilometer-wide path. Seven minutes later, the umbra reaches the Atlantic coast of Angola (05:57 UT) as it begins its trajectory across southern Africa (Figures 2 to 11). Quite coincidentally, some regions within the initial track through Angola were also in the path of the total solar eclipse of 2001 June 21 (Figure 2). The local residents there are indeed fortunate to witness two total eclipses of the Sun within the span of eighteen months.

The early morning eclipse lasts 51 seconds from the center line with the Sun 19° above the horizon. The umbra carves out a 50 kilometer-wide path as it sweeps across Angola in a southeastern direction. Briefly straddling the Angola/Zambia border (Figure 3), the shadow crosses eastern Namibia before entering northern Botswana (06:09 UT). The path width has grown to 60 kilometers and totality lasts 1 minute 11 seconds. Following the political boundary between Zimbabwe and Botswana, the umbra travels with a ground speed of 1.2 km/s. Bulawayo, Zimbabwe, lies just north of the track and its residents witness a deep partial eclipse of magnitude 0.987 at 06:14 UT.

The umbra crosses completely into Zimbabwe before entering northern South Africa at 06:19 UT (Figure 4). One minute later, the northern third of Kruger National Park is plunged into totality, which lasts 1 minute 25 seconds as the hidden Sun stands 42° above the horizon. Quickly crossing southern Mozambique, the shadow leaves the dark continent at 06:28 UT. The coastal city of Xai-Xai lies south of the center line but still experiences 58 seconds of totality before the umbra begins its long trek across the Indian Ocean.

The instant of greatest eclipse[1] occurs at 07:31:11 UT when the axis of the Moon's shadow passes closest to the center of Earth (gamma[2] = -0.302). The length of totality reaches its maximum duration of 2 minutes 4 seconds, the Sun's altitude is 72°, the path width is 87 kilometers and the umbra's velocity is 0.670 km/s. Unfortunately, the umbra is far at sea ~2000 kilometers southeast of Madagascar.

During the next hour and a half, no land is encountered as the eclipse track curves to the northeast and begins to narrow. In the final 90 seconds of its terrestrial trajectory, the umbra traverses South Australia (Figures 12 to 17). The coastal town of Ceduna lies at the center of the 35 kilometer-wide path. Totality lasts 33 seconds while the Sun stands 9° above the western horizon. The accelerating ground speed of the umbra already exceeds 5 km/s. In the remaining seconds, the increasingly elliptical shadow sweeps across 900 kilometers of the Australian Outback.

The umbra leaves Earth's surface at the sunset terminator at 09:12 UT. Over the course of 3 hours and 21 minutes, the Moon's umbra travels along a path approximately 12,000 kilometers long and covering 0.14% of Earth's surface area.

---

[1] The instant of greatest eclipse occurs when the distance between the Moon's shadow axis and Earth's geocenter reaches a minimum. Although greatest eclipse differs slightly from the instants of greatest magnitude and greatest duration (for total eclipses), the differences are usually quite small.

[2] Minimum distance of the Moon's shadow axis from Earth's center in units of equatorial Earth radii.

Maps of the Eclipse Path

Orthographic Projection Map of the Eclipse Path

Figure 1 is an orthographic projection map of Earth [adapted from Espenak, 1987] showing the path of penumbral (partial) and umbral (total) eclipse. The daylight terminator is plotted for the instant of greatest eclipse with north at the top. The sub-Earth point is centered over the point of greatest eclipse and is indicated with an asterisk-like symbol. The sub-solar point (Sun in zenith) at that instant is also shown.

The limits of the Moon's penumbral shadow define the region of visibility of the partial eclipse. This saddle-shaped region often covers more than half of Earth's daylight hemisphere and consists of several distinct zones or limits. At the northern and/or southern boundaries lie the limits of the penumbra's path. Partial eclipses have only one of these limits, as do central eclipses when the shadow axis falls no closer than about 0.45 radii from Earth's center. Great loops at the western and eastern extremes of the penumbra's path identify the areas where the eclipse begins/ends at sunrise and sunset, respectively. If the penumbra has both a northern and southern limit, the rising and setting curves form two separate, closed loops. Otherwise, the curves are connected in a distorted figure eight. Bisecting the 'eclipse begins/ends at sunrise and sunset' loops is the curve of maximum eclipse at sunrise (western loop) and sunset (eastern loop). The exterior tangency points **P1** and **P4** mark the coordinates where the penumbral shadow first contacts (partial eclipse begins) and last contacts (partial eclipse ends) Earth's surface. The path of the umbral shadow bisects the penumbral path from west to east and is shaded dark gray.

A curve of maximum eclipse is the locus of all points where the eclipse is at maximum at a given time. They are plotted at each half hour Universal Time (UT), and generally run from northern to southern penumbral limits, or from the maximum eclipse at sunrise or sunset curves to one of the limits. The outline of the umbral shadow is plotted every 10 minutes in UT. Curves of constant eclipse magnitude[3] delineate the locus of all points where the magnitude at maximum eclipse is constant. These curves run exclusively between the curves of maximum eclipse at sunrise and sunset. Furthermore, they are quasi-parallel to the northern/southern penumbral limits and the umbral paths of central eclipses. Northern and southern limits of the penumbra may be thought of as curves of constant magnitude of 0%, while adjacent curves are for magnitudes of 20%, 40%, 60% and 80%. The northern and southern limits of the path of total eclipse are curves of constant magnitude of 100%.

At the top of Figure 1, the Universal Time of geocentric conjunction between the Moon and Sun is given followed by the instant of greatest eclipse. The eclipse magnitude is given for greatest eclipse. For central eclipses (both total and annular), it is equivalent to the geocentric ratio of diameters of the Moon and Sun. Gamma is the minimum distance of the Moon's shadow axis from Earth's center in units of equatorial Earth radii. The shadow axis passes south of Earth's geocenter for negative values of Gamma. Finally, the Saros series number of the eclipse is given along with its relative sequence in the series.

Equidistant Conic Projection Map of the Eclipse Path

Figures 2, 3, 4 (covering Africa), 12 and 13 (covering Australia) are maps using an equidistant conic projection chosen to minimize distortion, and which isolate the African and Australian portions of the umbral path. Curves of maximum eclipse and constant eclipse magnitude are plotted and labeled at intervals ranging from 1 to 10 minutes, depending on the map scale and umbral shadow velocity. A linear scale is included for estimating approximate distances (kilometers). Within the northern and southern limits of the path of totality, the outline of the umbral shadow is plotted at intervals of 10 minutes or less. The duration of totality (minutes and seconds) and the Sun's altitude correspond to the local circumstances on the center line at each shadow position.

The scales used in the maps in these figures are as follows:

  Figure 2 —   1:31,804,000
  Figure 3 & 4 — 1:7,952,000
  Figure 12 —   1:20,240,000
  Figure 13 —   1:5,936,000

---

[3] Eclipse magnitude is defined as the fraction of the Sun's diameter occulted by the Moon. It is strictly a ratio of *diameters* and should not be confused with eclipse obscuration, which is a measure of the Sun's surface *area* occulted by the Moon. Eclipse magnitude may be expressed as either a percentage or a decimal fraction (e.g.: 50% or 0.50).

The positions of larger cities and metropolitan areas in and near the umbral path are depicted as black dots. The size of each city is logarithmically proportional to its population using 1990 census data (Rand McNally, 1991). City data from a geographic data base of over 90,000 positions are plotted to give as many locations as possible in the path of totality. Local circumstances have been calculated for many of these positions and can be found in Tables 11 through 18.

DETAILED MAPS OF THE UMBRAL PATH

The path of totality is plotted on a series detailed maps appearing in Figures 5 through 11 (Africa) and Figures 14 through 17 (Australia). The maps were chosen to isolate small regions along the entire land portion of the eclipse path. Curves of maximum eclipse are plotted at 1- or 2- minute intervals along the track and labeled with the center line duration of totality and the Sun's altitude. The maps are constructed from the Digital Chart of the World (DCW), a digital database of the world developed by the U. S. Defense Mapping Agency (DMA). The primary sources of information for the geographic database are the Operational Navigation Charts (ONC) and the Jet Navigation Charts (JNC) developed by the DMA.

The 1:2,000,000 scale of the eclipse maps is adequate for showing roads, villages and cities, required for eclipse expedition planning. Caution should be employed in using the maps since no distinction is made between major highways and second class soft-surface roads. Those who require more detailed maps of the eclipse track should plot the coordinates from Tables 7-9 on larger scale maps.

The DCW database was assembled in the 1980s and contains names of places that are no longer used in some parts of Africa, particularly Zimbabwe. Where possible, modern names have been substituted for those in the database but this correction could not be applied to all sites. Some areas of missing topographic data appear as blank or white rectangles on the map background.

Northern and southern limits as well as the center line of the path are plotted using data from Tables 7 and 8. Although no corrections have been made for center of figure or lunar limb profile, they have little or no effect at this scale. Atmospheric refraction has not been included, as it plays a significant role only at very low solar altitudes[4]. In any case, refraction corrections to the path are uncertain since they depend on the atmospheric temperature-pressure profile, which cannot be predicted in advance. If observations from the graze zones are planned, then the zones of grazing eclipse must be plotted on higher scale maps using coordinates in Table 9. See PLOTTING THE PATH ON MAPS for sources and more information. The paths also show the curves of maximum eclipse at 2-minute increments in UT. These maps are also available on the web at *http://sunearth.gsfc.nasa.gov/eclipse/TSE2002/TSE2002.html*).

ELEMENTS, SHADOW CONTACTS AND ECLIPSE PATH TABLES

The geocentric ephemeris for the Sun and Moon, various parameters, constants, and the Besselian elements (polynomial form) are given in Table 1. The eclipse elements and predictions were derived from the DE200 and LE200 ephemerides (solar and lunar, respectively) developed jointly by the Jet Propulsion Laboratory and the U. S. Naval Observatory for use in the *Astronomical Almanac* for 1984 and thereafter. Unless otherwise stated, all predictions are based on center of mass positions for the Moon and Sun with no corrections made for center of figure, lunar limb profile or atmospheric refraction. The predictions depart from normal IAU convention through the use of a smaller constant for the mean lunar radius k for all umbral contacts (see: LUNAR LIMB PROFILE). Times are expressed in either Terrestrial Dynamical Time (TDT) or in Universal Time (UT), where the best value of ΔT[5] available at the time of preparation is used.

From the polynomial form of the Besselian elements, any element can be evaluated for any time t_1 (in decimal hours) via the equation:

$$\mathbf{a} = a_0 + a_1 * t + a_2 * t^2 + a_3 * t^3 \quad (\text{or } \mathbf{a} = \Sigma\ [a_n * t^n]; n = 0 \text{ to } 3)$$

where: $\mathbf{a}$ = x, y, d, l_1, l_2, or μ
$\quad\quad\quad\ \ $ t = $t_1 - t_0$ (decimal hours) and t_0 = 8.00 TDT

---

[4] The primary effect of refraction is to shift the path opposite to that of the Sun's local azimuth. This amounts to approximately 0.5° at the extreme ends (i.e. - sunrise and sunset) of the umbral path.

[5] ΔT is the difference between Terrestrial Dynamical Time and Universal Time.

The polynomial Besselian elements were derived from a least-squares fit to elements rigorously calculated at five separate times over a 6-hour period centered at t_0. Thus, the equation and elements are valid over the period $5.00 \leq t_1 \leq 11.00$ TDT.

Table 2 lists all external and internal contacts of penumbral and umbral shadows with Earth. They include TDT times and geodetic coordinates with and without corrections for ΔT. The contacts are defined:

P1 - Instant of first external tangency of penumbral shadow cone with Earth's limb.
(partial eclipse begins)
P4 - Instant of last external tangency of penumbral shadow cone with Earth's limb.
(partial eclipse ends)
U1 - Instant of first external tangency of umbral shadow cone with Earth's limb.
(umbral eclipse begins)
U2 - Instant of first internal tangency of umbral shadow cone with Earth's limb.
U3 - Instant of last internal tangency of umbral shadow cone with Earth's limb.
U4 - Instant of last external tangency of umbral shadow cone with Earth's limb.
(umbral eclipse ends)

Similarly, the northern and southern extremes of the penumbral and umbral paths, and extreme limits of the umbral center line are given. The IAU (International Astronomical Union) longitude convention is used throughout this publication (i.e., for longitude, east is positive and west is negative; for latitude, north is positive and south is negative).

The path of the umbral shadow is delineated at 5-minute intervals in Universal Time in Table 3. Coordinates of the northern limit, the southern limit and the center line are listed to the nearest tenth of an arc-minute (~185 m at the Equator). The Sun's altitude, path width and umbral duration are calculated for the center line position. Table 4 presents a physical ephemeris for the umbral shadow at 5-minute intervals in UT. The center line coordinates are followed by the topocentric ratio of the apparent diameters of the Moon and Sun, the eclipse obscuration[6], and the Sun's altitude and azimuth at that instant. The central path width, the umbral shadow's major and minor axes and its instantaneous velocity with respect to Earth's surface are included. Finally, the center line duration of the umbral phase is given.

Local circumstances for each center line position listed in Tables 3 and 4 are presented in Table 5. The first three columns give the Universal Time of maximum eclipse, the center line duration of totality and the altitude of the Sun at that instant. The following columns list each of the four eclipse contact times followed by their related contact position angles and the corresponding altitude of the Sun. The four contacts identify significant stages in the progress of the eclipse. They are defined as follows:

First Contact — Instant of first external tangency between the Moon and Sun.
(partial eclipse begins)
Second Contact — Instant of first internal tangency between the Moon and Sun.
(central or umbral eclipse begins; total or annular eclipse begins)
Third Contact — Instant of last internal tangency between the Moon and Sun.
(central or umbral eclipse ends; total or annular eclipse ends)
Fourth Contact — Instant of last external tangency between the Moon and Sun.
(partial eclipse ends)

The position angles **P** and **V** identify the point along the Sun's disk where each contact occurs[7]. Second and third contact altitudes are omitted since they are always within 1° of the altitude at maximum eclipse.

Table 6 presents topocentric values from the central path at maximum eclipse for the Moon's horizontal parallax, semi-diameter, relative angular velocity with respect to the Sun, and libration in longitude. The altitude and azimuth of the Sun are given along with the azimuth of the umbral path. The northern limit position angle identifies the point on the lunar disk defining the umbral path's northern limit. It is measured counter-clockwise from the north point of the Moon. In addition, corrections to the path limits due to the lunar limb profile are listed. The irregular profile of the Moon results in a zone of 'grazing eclipse' at each limit that is delineated by interior and exterior contacts of lunar features with the Sun's limb. This geometry is described in greater detail in the section LIMB CORRECTIONS TO THE PATH LIMITS: GRAZE ZONES. Corrections to center line durations due to the lunar limb profile are also included. When added to the durations in Tables 3, 4, 5, 7 and 8, a slightly shorter central total phase is predicted along most of the path.

---

[6] Eclipse obscuration is defined as the fraction of the Sun's surface area occulted by the Moon.

[7] P is defined as the contact angle measured counter-clockwise from the *north* point of the Sun's disk. V is defined as the contact angle measured counter-clockwise from the *zenith* point of the Sun's disk.

To aid and assist in the plotting of the umbral path on large scale maps, the path coordinates are also tabulated at 30' intervals in longitude in Tables 7 and 8. The latitude of the northern limit, southern limit and center line for each longitude is tabulated to the nearest hundredth of an arc-minute (~18.5 m at the Equator) along with the Universal Time of maximum eclipse at each position. Finally, local circumstances on the center line at maximum eclipse are listed and include the Sun's altitude and azimuth, the umbral path width and the central duration of totality.

In applications where the zones of grazing eclipse are needed in greater detail, Tables 9 and 10 list these coordinates over land- based portions of the path at 30' intervals in longitude. The time of maximum eclipse is given at both northern and southern limits as well as the path's azimuth. The elevation and scale factors are also given (see: LIMB CORRECTIONS TO THE PATH LIMITS: GRAZE ZONES).

LOCAL CIRCUMSTANCES TABLES

Local circumstances for approximately 400 cities, metropolitan areas and places in Africa, Australia, etc., are presented in Tables 11 through 18. These tables give the local circumstances at each contact and at maximum eclipse[8] for every location. The coordinates are listed along with the location's elevation (meters) above sea-level, if known. If the elevation is unknown (i.e., not in the data base), then the local circumstances for that location are calculated at sea-level. In any case, the elevation does not play a significant role in the predictions unless the location is near the umbral path limits and the Sun's altitude is relatively small (<10°). The Universal Time of each contact is given to a tenth of a second, along with position angles **P** and **V** and the altitude of the Sun. The position angles identify the point along the Sun's disk where each contact occurs and are measured counter-clockwise (i.e., eastward) from the north and zenith points, respectively. Locations outside the umbral path miss the umbral eclipse and only witness first and fourth contacts. The Universal Time of maximum eclipse (either partial or total) is listed to a tenth of a second. Next, the position angles **P** and **V** of the Moon's disk with respect to the Sun are given, followed by the altitude and azimuth of the Sun at maximum eclipse. Finally, the corresponding eclipse magnitude and obscuration are listed. For umbral eclipses (both annular and total), the eclipse magnitude is identical to the topocentric ratio of the Moon's and Sun's apparent diameters.

Two additional columns are included if the location lies within the path of the Moon's umbral shadow. The **umbral depth** is a relative measure of a location's position with respect to the center line and path limits. It is a unitless parameter which is defined as:

$$\mathbf{u} = 1 - \text{abs}(\mathbf{x}/\mathbf{R}) \qquad [1]$$

where: $\mathbf{u}$ = umbral depth
$\mathbf{x}$ = perpendicular distance from the shadow axis (kilometers)
$\mathbf{R}$ = radius of the umbral shadow as it intersects Earth's surface (kilometers)

The umbral depth for a location varies from 0.0 to 1.0. A position at the path limits corresponds to a value of 0.0 while a position on the center line has a value of 1.0. The parameter can be used to quickly determine the corresponding center line duration. Thus, it is a useful tool for evaluating the trade-off in duration of a location's position relative to the center line. Using the location's duration and umbral depth, the center line duration is calculated as:

$$\mathbf{D} = \mathbf{d} / (1 - (1 - \mathbf{u})^2)^{1/2} \text{ seconds} \qquad [2]$$

where: $\mathbf{D}$ = duration of totality on the center line (seconds)
$\mathbf{d}$ = duration of totality at location (seconds)
$\mathbf{u}$ = umbral depth

The final column gives the duration of totality. The effects of refraction have not been included in these calculations, nor have there been any corrections for center of figure or the lunar limb profile.

Locations were chosen based on general geographic distribution, population, and proximity to the path. The primary source for geographic coordinates is *The New International Atlas* (Rand McNally, 1991). Elevations for major cities were taken from *Climates of the World* (U. S. Dept. of Commerce, 1972). In this rapidly changing political world, it is often difficult to ascertain the correct name or spelling for a given location. Therefore, the information presented here is

---

[8] For partial eclipses, maximum eclipse is the instant when the greatest fraction of the Sun's diameter is occulted. For total eclipses, maximum eclipse is the instant of mid-totality.

for location purposes only and is not meant to be authoritative. Furthermore, it does not imply recognition of status of any location by the United States Government. We hereby solicit corrections to names, spellings, coordinates and elevations in order to update the geographic data base for future eclipse predictions.

For countries in the path of totality, expanded versions of the local circumstances tables listing many more locations are available via a special web site of supplemental material for the total solar eclipse of 2002 (*http://sunearth.gsfc.nasa.gov/eclipse/TSE2002/TSE2002.html*).

ESTIMATING TIMES OF SECOND AND THIRD CONTACTS

The times of second and third contact for any location not listed in this publication can be estimated using the detailed maps (Figures 5-11, and 14-17). Alternatively, the contact times can be estimated from maps on which the umbral path has been plotted. Tables 7 and 8 list the path coordinates conveniently arranged in 30' increments of longitude to assist plotting by hand. The path coordinates in Table 3 define a line of maximum eclipse at 5-minute increments in time. These lines of maximum eclipse each represent the projection diameter of the umbral shadow at the given time. Thus, any point on one of these lines will witness maximum eclipse (i.e., mid-totality) at the same instant. The coordinates in Table 3 should be plotted on the map in order to construct lines of maximum eclipse.

The estimation of contact times for any one point begins with an interpolation for the time of maximum eclipse at that location. The time of maximum eclipse is proportional to a point's distance between two adjacent lines of maximum eclipse, measured along a line parallel to the center line. This relationship is valid along most of the path with the exception of the extreme ends, where the shadow experiences its largest acceleration. The center line duration of totality **D** and the path width **W** are similarly interpolated from the values of the adjacent lines of maximum eclipse as listed in Table 3. Since the location of interest probably does not lie on the center line, it is useful to have an expression for calculating the duration of totality **d** as a function of its perpendicular distance **a** from the center line:

$$\mathbf{d} = \mathbf{D}\,(1 - (2\,\mathbf{a}/\mathbf{W})^2)^{1/2} \text{ seconds} \qquad [3]$$

where: **d** = duration of totality at desired location (seconds)
D = duration of totality on the center line (seconds)
a = perpendicular distance from the center line (kilometers)
W = width of the path (kilometers)

If t_m is the interpolated time of maximum eclipse for the location, then the approximate times of second and third contacts (t_2 and t_3, respectively) are:

Second Contact: $\quad t_2 = t_m - d/2 \qquad [4]$
Third Contact: $\quad t_3 = t_m + d/2 \qquad [5]$

The position angles of second and third contact (either **P** or **V**) for any location off the center line are also useful in some applications. First, linearly interpolate the center line position angles of second and third contacts from the values of the adjacent lines of maximum eclipse as listed in Table 5. If X_2 and X_3 are the interpolated center line position angles of second and third contacts, then the position angles x_2 and x_3 of those contacts for an observer located **a** kilometers from the center line are:

Second Contact: $\quad x_2 = X_2 - \arcsin(2\,\mathbf{a}/\mathbf{W}) \qquad [6]$
Third Contact: $\quad x_3 = X_3 + \arcsin(2\,\mathbf{a}/\mathbf{W}) \qquad [7]$

where: x_n = interpolated position angle (either **P** or **V**) of contact **n** at location
X_n = interpolated position angle (either **P** or **V**) of contact **n** on center line
a = perpendicular distance from the center line (kilometers)
(use negative values for locations south of the center line)
W = width of the path (kilometers)

MEAN LUNAR RADIUS

A fundamental parameter used in eclipse predictions is the Moon's radius k, expressed in units of Earth's equatorial radius. The Moon's actual radius varies as a function of position angle and libration due to the irregularity in the limb profile. From 1968 through 1980, the Nautical Almanac Office used two separate values for k in their predictions. The larger value (k=0.2724880), representing a mean over topographic features, was used for all penumbral (exterior) contacts and for annular eclipses. A smaller value (k=0.272281), representing a mean minimum radius, was reserved exclusively for umbral (interior) contact calculations of total eclipses [*Explanatory Supplement*, 1974]. Unfortunately, the use of two different values of k for umbral eclipses introduces a discontinuity in the case of hybrid or annular-total eclipses.

In August 1982, the International Astronomical Union (IAU) General Assembly adopted a value of k=0.2725076 for the mean lunar radius. This value is now used by the Nautical Almanac Office for all solar eclipse predictions [Fiala and Lukac, 1983] and is currently the best mean radius, averaging mountain peaks and low valleys along the Moon's rugged limb. The adoption of one single value for k eliminates the discontinuity in the case of annular-total eclipses and ends confusion arising from the use of two different values. However, the use of even the best 'mean' value for the Moon's radius introduces a problem in predicting the true character and duration of umbral eclipses, particularly total eclipses. A total eclipse can be defined as an eclipse in which the Sun's disk is completely occulted by the Moon. This cannot occur so long as any photospheric rays are visible through deep valleys along the Moon's limb [Meeus, Grosjean and Vanderleen, 1966]. But the use of the IAU's mean k guarantees that some annular or annular-total eclipses will be misidentified as total. A case in point is the eclipse of 3 October 1986. Using the IAU value for k, the *Astronomical Almanac* identified this event as a total eclipse of 3 seconds duration when it was, in fact, a beaded annular eclipse. Since a smaller value of k is more representative of the deeper lunar valleys and hence the minimum solid disk radius, it helps ensure the correct identification of an eclipse's true nature.

Of primary interest to most observers are the times when umbral eclipse begins and ends (second and third contacts, respectively) and the duration of the umbral phase. When the IAU's value for k is used to calculate these times, they must be corrected to accommodate low valleys (total) or high mountains (annular) along the Moon's limb. The calculation of these corrections is not trivial but must be performed, especially if one plans to observe near the path limits [Herald, 1983]. For observers near the center line of a total eclipse, the limb corrections can be more closely approximated by using a smaller value of k which accounts for the valleys along the profile.

This publication uses the IAU's accepted value of k=0.2725076 for all penumbral (exterior) contacts. In order to avoid eclipse type misidentification and to predict central durations which are closer to the actual durations at total eclipses, we depart from standard convention by adopting the smaller value of k=0.272281 for all umbral (interior) contacts. This is consistent with predictions in *Fifty Year Canon of Solar Eclipses: 1986 - 2035* [Espenak, 1987]. Consequently, the smaller k produces shorter umbral durations and narrower paths for total eclipses when compared with calculations using the IAU value for k. Similarly, predictions using a smaller k result in longer umbral durations and wider paths for annular eclipses than do predictions using the IAU's k.

LUNAR LIMB PROFILE

Eclipse contact times, magnitude and duration of totality all depend on the angular diameters and relative velocities of the Moon and Sun. Unfortunately, these calculations are limited in accuracy by the departure of the Moon's limb from a perfectly circular figure. The Moon's surface exhibits a rather dramatic topography, which manifests itself as an irregular limb when seen in profile. Most eclipse calculations assume some mean radius that averages high mountain peaks and low valleys along the Moon's rugged limb. Such an approximation is acceptable for many applications, but if higher accuracy is needed, the Moon's actual limb profile must be considered. Fortunately, an extensive body of knowledge exists on this subject in the form of Watts' limb charts [Watts, 1963]. These data are the product of a photographic survey of the marginal zone of the Moon and give limb profile heights with respect to an adopted smooth reference surface (or datum). Analyses of lunar occultations of stars by Van Flandern [1970] and Morrison [1979] have shown that the average cross-section of Watts' datum is slightly elliptical rather than circular. Furthermore, the implicit center of the datum (i.e., the center of figure) is displaced from the Moon's center of mass. In a follow-up analysis of 66,000 occultations, Morrison and Appleby [1981] have found that the radius of the datum appears to vary with libration. These variations produce systematic errors in Watts' original limb profile heights that attain 0.4 arc-seconds at some position angles. Thus, corrections to Watts' limb data are necessary to ensure that the reference datum is a sphere with its center at the center of mass.

The Watts charts were digitized by Her Majesty's Nautical Almanac Office in Herstmonceux, England, and transformed to grid-profile format at the U. S. Naval Observatory. In this computer readable form, the Watts limb charts lend themselves to the generation of limb profiles for any lunar libration. Ellipticity and libration corrections may be applied to refer the profile to the Moon's center of mass. Such a profile can then be used to correct eclipse predictions which have been generated using a mean lunar limb.

Along the path, the Moon's topocentric libration (physical + optical) in longitude ranges from l=+4.5° to l=+2.9°. Thus, a limb profile with the appropriate libration is required in any detailed analysis of contact times, central durations, etc. But a profile with an intermediate value is useful for planning purposes and may even be adequate for most applications. The lunar limb profile presented in Figure 18 includes corrections for center of mass and ellipticity [Morrison and Appleby, 1981]. It is generated for 06:15 UT, which corresponds to western Zimbabwe near the border with Botswana. The Moon's topocentric libration is l=+4.37°, and the topocentric semi-diameters of the Sun and Moon are 973.7 and 992.2 arc-seconds, respectively. The Moon's angular velocity with respect to the Sun is 0.468 arc-seconds per second.

The radial scale of the limb profile in Figure 18 (at bottom) is greatly exaggerated so that the true limb's departure from the mean lunar limb is readily apparent. The mean limb with respect to the center of figure of Watts' original data is shown (dashed) along with the mean limb with respect to the center of mass (solid). Note that all the predictions presented in this publication are calculated with respect to the latter limb unless otherwise noted. Position angles of various lunar features can be read using the protractor marks along the Moon's mean limb (center of mass). The position angles of second and third contact are clearly marked along with the north pole of the Moon's axis of rotation and the observer's zenith at mid-totality. The dashed line with arrows at either end identifies the contact points on the limb corresponding to the northern and southern limits of the path. To the upper left of the profile are the Sun's topocentric coordinates at maximum eclipse. They include the right ascension *R.A.*, declination *Dec.*, semi-diameter *S.D.* and horizontal parallax *H.P.* The corresponding topocentric coordinates for the Moon are to the upper right. Below and left of the profile are the geographic coordinates of the center line at 06:15 UT while the times of the four eclipse contacts at that location appear to the lower right. Directly below the profile are the local circumstances at maximum eclipse. They include the Sun's altitude and azimuth, the path width, and central duration. The position angle of the path's northern/southern limit axis is *PA(N.Limit)* and the angular velocity of the Moon with respect to the Sun is *A.Vel.(M:S)*. At the bottom left are a number of parameters used in the predictions, and the topocentric lunar librations appear at the lower right.

In investigations where accurate contact times are needed, the lunar limb profile can be used to correct the nominal or mean limb predictions. For any given position angle, there will be a high mountain (annular eclipses) or a low valley (total eclipses) in the vicinity that ultimately determines the true instant of contact. The difference, in time, between the Sun's position when tangent to the contact point on the mean limb and tangent to the highest mountain (annular) or lowest valley (total) at actual contact is the desired correction to the predicted contact time. On the exaggerated radial scale of Figure 18, the Sun's limb can be represented as an epicyclic curve that is tangent to the mean lunar limb at the point of contact and departs from the limb by **h** through:

$$\mathbf{h} = \mathbf{S}\,(\mathbf{m}-1)\,(1-\cos[\mathbf{C}]) \qquad [8]$$

where: **h** = departure of Sun's limb from mean lunar limb
S = Sun's semi-diameter
m = eclipse magnitude
C = angle from the point of contact

Herald [1983] has taken advantage of this geometry to develop a graphical procedure for estimating correction times over a range of position angles. Briefly, a displacement curve of the Sun's limb is constructed on a transparent overlay by way of equation [8]. For a given position angle, the solar limb overlay is moved radially from the mean lunar limb contact point until it is tangent to the lowest lunar profile feature in the vicinity. The solar limb's distance **d** (arc-seconds) from the mean lunar limb is then converted to a time correction Δ by:

$$\Delta = \mathbf{d}\,\mathbf{v}\,\cos[\mathbf{X} - \mathbf{C}] \qquad [9]$$

where: Δ = correction to contact time (seconds)
d = distance of Solar limb from Moon's mean limb (arc-sec)
v = angular velocity of the Moon with respect to the Sun (arc-sec/sec)
X = center line position angle of the contact
C = angle from the point of contact

This operation may be used for predicting the formation and location of Baily's beads. When calculations are performed over a large range of position angles, a contact time correction curve can then be constructed.

Since the limb profile data are available in digital form, an analytical solution to the problem is possible that is quite straightforward and robust. Curves of corrections to the times of second and third contact for most position angles have been computer generated and are plotted in Figure 18. The circular protractor scale at the center represents the nominal contact time using a mean lunar limb. The departure of the contact correction curves from this scale graphically illustrates the time correction to the mean predictions for any position angle as a result of the Moon's true limb profile. Time corrections external to the circular scale are added to the mean contact time; time corrections internal to the protractor are subtracted from the mean contact time. The magnitude of the time correction at a given position angle is measured using any of the four radial scales plotted at each cardinal point.

For example, Table 16 gives the following data for Beitbridge, Zimbabwe:

   Second Contact = 06:18:03.4 UT P_2=127°
   Third Contact   = 06:19:25.0 UT P_3=286°

Using Figure 18, the measured time corrections and the resulting contact times are:

   C_2=–2.2 seconds;   Second Contact = 06:18:03.4—2.2s = 06:18:01.2 UT
   C_3=–2.7 seconds;   Third Contact = 06:19:25.0 –2.7s = 06:19:22.3 UT

The above corrected values are within 0.1 seconds of a rigorous calculation using the true limb profile.

Lunar limb profile diagrams for several other positions/times along the path of totality are available via a special web site of supplemental material for the total solar eclipse of 2002 (*http://sunearth.gsfc.nasa.gov/eclipse/TSE2002/TSE2002.html*).

LIMB CORRECTIONS TO THE PATH LIMITS: GRAZE ZONES

The northern and southern umbral limits provided in this publication were derived using the Moon's center of mass and a mean lunar radius. They have not been corrected for the Moon's center of figure or the effects of the lunar limb profile. In applications where precise limits are required, Watts' limb data must be used to correct the nominal or mean path. Unfortunately, a single correction at each limit is not possible since the Moon's libration in longitude and the contact points of the limits along the Moon's limb each vary as a function of time and position along the umbral path. This makes it necessary to calculate a unique correction to the limits at each point along the path. Furthermore, the northern and southern limits of the umbral path are actually paralleled by a relatively narrow zone where the eclipse is neither penumbral nor umbral. An observer positioned here will witness a slender solar crescent that is fragmented into a series of bright beads and short segments whose morphology changes quickly with the rapidly varying geometry between the limbs of the Moon and the Sun. These beading phenomena are caused by the appearance of photospheric rays that alternately pass through deep lunar valleys and hide behind high mountain peaks as the Moon's irregular limb grazes the edge of the Sun's disk. The geometry is directly analogous to the case of grazing occultations of stars by the Moon. The graze zone is typically 5 to 10 kilometers wide and its interior and exterior boundaries can be predicted using the lunar limb profile. The interior boundaries define the actual limits of the umbral eclipse (both total and annular) while the exterior boundaries set the outer limits of the grazing eclipse zone.

Table 6 provides topocentric data and corrections to the path limits due to the true lunar limb profile. At 5-minute intervals, the table lists the Moon's topocentric horizontal parallax, semi-diameter, relative angular velocity of the Moon with respect to the Sun and lunar libration in longitude. The Sun's center line altitude and azimuth is given, followed by the azimuth of the umbral path. The position angle of the point on the Moon's limb which defines the northern limit of the path is measured counter-clockwise (i.e., eastward) from the north point on the limb. The path corrections to the northern and southern limits are listed as interior and exterior components in order to define the graze zone. Positive corrections are in the northern sense while negative shifts are in the southern sense. These corrections (minutes of arc in latitude) may be added directly to the path coordinates listed in Table 3. Corrections to the center line umbral durations due to the lunar limb profile are also included and they are all negative. Thus, when added to the central durations given in Tables 3, 4, 5, 7 and 8, a slightly shorter central total phase is predicted.

Detailed coordinates for the zones of grazing eclipse at each limit for all land-based sections of the path are presented in Tables 9 and 10. Given the uncertainties in the Watts data, these predictions should be accurate to ±0.3 arc-seconds. The interior graze coordinates take into account the deepest valleys along the Moon's limb which produce the simultaneous second and third contacts at the path limits. Thus, the interior coordinates define the true edge of the path of totality. They are calculated from an algorithm which searches the path limits for the extreme positions where no photospheric beads are visible along a ±30° segment of the Moon's limb, symmetric about the extreme contact points at the instant of maximum eclipse. The exterior graze coordinates are somewhat arbitrarily defined and calculated for the geodetic positions where an unbroken photospheric crescent of 60° in angular extent is visible at maximum eclipse.

In Tables 9 and 10, the graze zone latitudes are listed every 30' in longitude (at sea level) and include the time of maximum eclipse at the northern and southern limits as well as the path's azimuth. To correct the path for locations above sea level, *Elev Fact*[9] is a multiplicative factor by which the path must be shifted north perpendicular to itself (i.e., perpendicular to path azimuth) for each unit of elevation (height) above sea level. To calculate the shift, a location's elevation is multiplied by the *Elev Fact* value. Positive values (usually the case for eclipses in the Southern Hemisphere) indicate that the path must be shifted north. For instance, if one's elevation is 1000 meters above sea level and the *Elev Fact* value is +0.50, then the shift is +500m (= 1000m x +0.50). Thus, the observer must shift the path coordinates 500 meters in a direction perpendicular to the path and in a positive or northerly sense.

The final column of Tables 9 and 10 list the *Scale Fact* (km/arc-second). This scaling factor provides an indication of the width of the zone of grazing phenomena, due to the topocentric distance of the Moon and the projection geometry of the Moon's shadow on Earth's surface. Since the solar chromosphere has an apparent thickness of about 3 arc-seconds, and assuming a *Scale Fact* value of 2 km/arc-seconds, then the chromosphere should be visible continuously during totality for any observer in the path who is within 6 kilometers (=2x3) of each interior limit. However, the most dynamic beading phenomena occurs within 1.5 arc-seconds of the Moon's limb. Using the above Scale Factor, this translates into the first 3 kilometers inside the interior limits. But observers should position themselves at least 1 kilometer inside the interior limits (south of the northern interior limit or north of the southern interior limit) in order to ensure that they are inside the path due of to small uncertainties in Watts' data and the actual path limits.

For applications where the zones of grazing eclipse are needed at a higher frequency in longitude interval, tables of coordinates every 7.5' in longitude are available via a special web site for the total solar eclipse of 2002 (*http://sunearth.gsfc.nasa.gov/eclipse/TSE2002/TSE2002.html*).

SAROS HISTORY

The periodicity and recurrence of solar (and lunar) eclipses is governed by the Saros cycle, a period of approximately 6,585.3 days (18 years 11 days 8 hours). When two eclipses are separated by a period of one Saros, they share a very similar geometry. The eclipses occur at the same node with the Moon at nearly the same distance from Earth and at the same time of year. Thus, the Saros is useful for organizing eclipses into families or series. Each series typically lasts 12 to 13 centuries and contains 70 or more eclipses.

The total eclipse of 2002 December 04 is the twenty-second member of Saros series 142 (Table 19), as defined by van den Bergh [1955]. All eclipses in the series occur at the Moon's descending node and the Moon moves northward with each member in the family (i.e. - gamma[10] increases). Saros 142 is a young series which began with a small partial eclipse at high southern latitudes on 1624 Apr 17. After seven partial eclipses each of increasing magnitude, the first umbral eclipse occurred on 1750 Jul 03. This unusual annular eclipse was non-central and had no southern limit as the northern edge of the Moon's antumbral shadow briefly grazed Earth in the South Pacific. One saros later, the eclipse of 1768 Jul 14 was a short hybrid eclipse just south of Australia. The next eclipse in the series (1786 Jul 25) began a long sequence of total eclipses. The maximum duration of totality was just under 1 minute and the path passed through South Africa.

Over the next two centuries, the duration of totality of each eclipse has gradually increased as the Moon moved closer to perigee with every succeeding event. However, during the twentieth century, the maximum duration has hovered near 2 minutes as Earth approaches perihelion. In the next century, the duration will continue to increase as the umbral shadow passes progressively closer to Earth's geocenter. The most recent eclipse of the series occurred on 1984 Nov 22. The umbral path ran through Papua New Guinea and the South Pacific. After 2002, the following member occurs on 2020 Dec 14. Its track stretches from the South Pacific, across South America and into the South Atlantic *(http://sunearth.gsfc.nasa.gov/eclipse/eclipse/SEplot/SE2002Dec14T.gif)*.

---

[9] The elevation factor is the product, $\tan(90-A) * \sin(D)$, where A is the altitude of the Sun and D is the difference between the azimuth of the Sun and the azimuth of the limit line, with the sign selected to be positive if the path should be shifted north with positive elevations above sea level.

[10] Minimum distance of the Moon's shadow axis from Earth's center in units of equatorial Earth radii. Gamma defines the instant of greatest eclipse and takes on negative values south of the Earth's center

By the twenty-third century, Saros 142 will be producing total eclipses with maximum durations exceeding 6 minutes in the northern tropics. The longest eclipse of the series occurs on 2291 May 28 and will last 6 minutes 34 seconds. The path of each succeeding event swings further north and a 6-minute eclipse will be visible from much of the United States on 2345 Jun 30. The duration of totality steadily decreases as the eclipse paths run through higher northern latitudes. The last total eclipse occurs in northern Canada on 2543 Oct 29. Although the event is central and lasts over 2 minutes, the northern edge of the umbral shadow will miss Earth completely. Partial eclipses will be visible from the northern hemisphere for the next three centuries. Saros 142 reaches its conclusion with the partial eclipse of 2886 May 25. A detailed list of eclipses in Saros series 142 appears in Table 19. For a more detailed list including local circumstances at greatest eclipse, see: *http://sunearth.gsfc.nasa.gov/eclipse/SEsaros/SEsaros142.html*

In summary, Saros series 142 includes 72 eclipses. It begins with 7 partials, followed by 1 annular, 1 hybrid, 43 total eclipses and finally ends with 20 more partials. The total duration of Saros 142 is 1280.1 years.

Weather Prospects for the Eclipse

Introduction

Just eighteen months after the 2001 eclipse the Moon's shadow returns to southern Africa. The shift in season marks a dramatic change in the weather, for where 2001's eclipse occurred during the dry southern winter, that of 2002 is in the midst of the wet summer. This time around, Australia offers a tempting alternative to the African rainy season, but the choices presented to the observer will be difficult to reconcile. In Australia, better weather prospects come at the cost of a short eclipse and a very low solar altitude that will magnify the effects of any cloud that might be present.

African Weather Prospects

African Overview

Africa's most significant climatic feature is its variable rainfall, driven by the continent's position between Earth's anticyclonic belts and its location astride the equator. In the southern part of the continent, the rainfall is highly seasonal, with a pronounced dry season in the winter (June) and an equally prominent wet season in the summer. December (start of southern hemisphere summer) is well into the rainy season, though not the wettest month, and eclipse watchers must contend with a very different climate from that which occurred for the 2001 eclipse.

In December, tropical moisture pushes well southward as high temperatures under an overhead Sun drag Earth's weather equator, the Inter-tropical Convergence Zone (ITCZ), into northern Zambia and central Mozambique (Figure 19). The Indian Ocean air behind the ITCZ is a generous reservoir of moisture, always ready to push southward when opportunity permits, but normally held at bay by the easterly trade winds. A second source of sub-tropical moisture can be found on the Atlantic side of the continent, where west and northwest winds cover Congo, Angola, and northwest Zambia with a moist airmass in December. This airmass is also opposed by the easterly trade winds; the boundary between the two is known as the Inter-Ocean Convergence Zone (IOCZ).

The anticyclonic (high-pressure) belt can be found, on average, along the 35th parallel, just south of the tip of the continent. The mean pressure pattern for December (Figure 19) shows anticyclones on both the Atlantic and Indian Ocean sides of the continent with a weak low pressure trough running north-to-south over Angola, Zambia and Botswana (the Botswana low). The images suggests that they are relatively static, but daily weather charts show that they move from west to east in a semi-regular cycle, bringing an ever-changing pattern of cloud and sun. The anticyclones are of critical importance to the eclipse chaser for they are the main controllers of sunny skies, bringing the dry trade winds from the Indian Ocean onto the continent.

Overlying the anticyclones is a zone of westerly winds that carry weather disturbances of their own. These upper level troughs and lows have a great influence on the weather below as they destabilize the air column and allow convective showers and thunderstorms to grow much more readily than in a stable environment.

Moist air from the northwest and dry air from the southeast moderate the changeable weather and cloudiness across southern Africa in December. Under the influence of passing anticyclones, waxing and waning northerly winds, upper level disturbances, occasional frontal systems, and winds deflected by mountainous terrain, a dynamic climatology of cloud and sun competes for the eclipse watcher's attention. But in spite of the variability of the weather systems there is a generous supply of sunny weather in December in southern Africa to attract the eclipse-watcher. The best areas are in northern South Africa and southeast Zimbabwe. Should movement be necessary as eclipse day approaches, the many countries that are spread out along the track will complicate last-minute adjustments in observing sites, but the main controls on the weather should be predictable for several days in advance.

The Countryside

For Angola, this is the second eclipse in as many years, and for a few small coastal communities north of Benguela, a second chance to see totality. The two tracks actually cross a hundred kilometers offshore, but are close enough on the coast that a very narrow strip defined by the south limit of the eclipse of 2001 and the north limit of 2002 will enjoy a rerun of the spectacle. Beyond this interesting coincidence however, Angola has little to offer the eclipse-seeker because of its prolonged civil war. Recent events have seen an effective counter-offensive by government troops, but the area of fighting is traversed by the eclipse track and millions of land mines make travel very hazardous.

Leaving Angola, the Moon's shadow skips along the Zambian border and then crosses a narrow extension of Namibia known as the Caprivi Strip before heading southeastward into Botswana. The Caprivi landscape features an expansive broad-leafed forest that bespeaks of its relatively generous precipitation and several prominent rivers, particularly the Okavango. The eclipse path lies across East Caprivi, about 40 km west of the town of Kongola. The track is easily reached by the Golden Highway between Kongola and Katima Mulilo. The narrowness of the Caprivi Strip and the lack of north-south roads limit the ability to move to clearer skies if the day is cloudy.

In Botswana, the track crosses the marvelous Chobe Game Park but the wet season severely restricts travel within the Park. December is still early in the rainy season and the eclipse may be accessible with four-wheel-drive vehicles, but it is likely easier (though farther) to travel from Victoria Falls or Livingstone, Zambia, through Zambia to Sesheke and view the eclipse from the Caprivi Strip. Farther south, the track zigzags back and forth between Botswana and Zimbabwe, passing through Hwange National Park, one of the finest conservation areas in the world. The center line within the park is relatively remote and areas to the southeast, closer to Bulawayo where the road network is denser, will be more accessible.

In southern Zimbabwe, the shadow path moves into better weather prospects. Between Plumtree and Beitbridge the center line can be followed for nearly 400 km on a network of roads, allowing a generous selection of eclipse sites and easy movement in case of adverse weather. Bulawayo lies only 50 km outside the north limit. From this city, major roads lead southwest to Plumtree and southeast to Beitbridge, both of which lie very nearly on the centerline. Access to this area is also possible from South Africa but the deteriorating political situation in Zimbabwe should be carefully reevaluated, especially in the weeks leading up to the eclipse.

Leaving Beitbridge, the eclipse track crosses briefly into South Africa, passing through the Venda Region and the northern part of Kruger National Park. The terrain becomes much rougher than the other parts of the eclipse track, crossing the peaks of the Soutpansberg, a mysterious land of magical lakes, rock engravings, legend and beauty. The Soutpansberg is richly vegetated, in contrast to the low veld over Zimbabwe, but the greenery is also a sign of the frequent mists and rains.

Kruger National Park is likely to be one of the primary destinations for the 2002 eclipse. A jewel of South Africa, the park is regarded as one of the finest wildlife management areas in the world. The center line travels across the drier part of the park, though in December the vegetation is a lush green because of the summer rains. Punda Maria and Shingwedzi are the two major camps within the zone of totality; Shingwedzi is barely north of the centerline. Of all of the National Parks crossed by this eclipse, Kruger is the easiest to reach and provides the greatest range of facilities.

Leaving South Africa, the track descends onto the coastal lowlands of Mozambique. Here, access to the center line can be obtained by following the coastal highway northward from the capital Maputo, crossing the Limpopo River near Xai-Xai (Shy-Shy) and proceeding onward another 40 km. The region has numerous palm-fringed beaches and resorts that should provide convenient bases for eclipse chasers. Floods in 2000 severely damaged roads and bridges in the Xai-Xai area though most of the route from Maputo to the center line was left in relatively good condition.

Malaria is a severe problem in southern Africa during the wet season and no part of the eclipse track is immune from the threat. The disease is becoming increasingly resistant to drugs and stringent precautions should be taken by all eclipse observers to minimize the chances of being infected. Light-colored clothing, mosquito repellants, and an appropriate drug regimen will help reduce the risk. Most African hotel beds are provided with mosquito netting for security at night. With good medical advice and sensible precautions, the risk of malaria can be minimized.

LARGE SCALE WEATHER PATTERNS

The eclipse track is squeezed between the moist air circulations in the north and the drier trade winds in the south. At its early part, in Angola, the track is affected by the Atlantic monsoon winds, but farther southeastward the path moves gradually into the drier trade wind climate as it approaches the Indian Ocean. The presence of the easterly trades is no guarantee of fine weather, however, as the controlling anticyclones have their own embedded disturbances that team up with the northern winds to bring the seasonal rains that define December's weather.

Most of the eclipse track is found in the zone of easterly trade winds. These winds are caused by the circulation around the belt of high-pressure anticyclones located south of the continent; the dry air in the center of each anticyclone is a virtual guarantee of sunny skies. Unfortunately, each high is separated from the other by a trough of lower pressure that usually contains the remnants of an ocean cold front and brings cloudy, wet weather if sufficient moisture is present. The eastward trek of the anticyclones and the intervening troughs brings an alternating pattern of rain and sun to South Africa. Secondary highs and high-pressure ridges also complicate the weather pattern, frequently reaching northward to extend the anticyclones' influence to the eclipse track.

The boundaries between Congo, Indian Ocean, and anticyclone airmasses create zones of wind convergence where air is forced upward. Since rising air cools and becomes moist, the boundaries are marked by extensive cloudiness and precipitation, usually in the form of thunderstorms. These convergence zones are not fixed in position, but slosh back and forth under the influence of high and low pressure systems as they pass across the continent. In general northerly flows are moist and southerly are dry.

One of the major controls on the weather across the eclipse track is a persistent heat low that forms over Botswana in the summer months. The circulation around this low brings northerly winds to the mid-continent and has the effect of pushing moist and unstable Angolan and Indian Ocean air southward into Botswana, Mozambique and northern South Africa when it intensifies. The moisture is usually capped by a temperature inversion that suppresses the development of convective clouds, but when upper level troughs or lows are also present, much more instability is created and the airmass erupts in showers and thunderstorms. These bring much-needed rain to the area, but also form extensive cloud shields that can spread across much of the eclipse track.

As in all climatologies, Africa has its share of typical weather patterns. Recognition of these patterns could help in establishing an eclipse site where sunshine is most likely.

One of the more effective rain-producing patterns begins with a high-pressure system located over the eastern eclipse track, usually with sunny skies overhead. On its western side, satellite images often show the formation of a long band of cloud angling from coastal Angola across Namibia and Botswana into central South Africa. This cloudiness is associated with upper level changes in the winds that occur near the Botswana low. The cloud band in the satellite images shows that the cloud is linked to the tropical moisture behind the IOCZ. A generous rainfall is produced if the Botswana low strengthens along this convergence boundary.

Southwest of the continent, some hundreds of kilometers distant, a band of frontal cloudiness marks the approach of an inter-anticyclonic trough. The approach of this trailing system induces the Botswana cloud band to move eastward to replace the sunny skies over South Africa's northern provinces, ending good eclipse weather. Occasionally the frontal cloud band will join with the leading cloud band, but the effects of the union are usually felt over southern regions, well away from the eclipse track.

Another significant pattern develops when an anticyclone invades the eastern parts of South Africa or strengthens in the waters offshore while a similar ridge prevails along the Atlantic coast. Between the two highs, the Botswana heat low deepens and stronger northerly winds develop on its east side. These northerlies carry high humidity tropical air southward into South Africa where mountainous terrain and instability can induce flooding rains. The presence of the high along the Indian Ocean coast often prevents this rainfall from moving into the eastern parts of the eclipse track, at least in the early stages of the event. In many respects this pattern is similar to the first example above, except that the intensity is more pronounced and the area of precipitation is likely narrower, in spite of higher rainfall amounts. A major difference, however, is that this second pattern tends to linger in place rather than moving eastward.

A third pattern begins when a cold low forms in the upper atmosphere, typically over the west coast of South Africa from where it drifts slowly eastward over a period of several days, isolated from the usual westerly flow aloft. The low induces a wide assortment of unfavorable weather to much of southern Africa. Moist tropical air flows southward across the mid continent, blocked from eastward motion only by the western slopes of the Drakensberg Mountains, bringing widespread rains. Shortly thereafter, and on the opposite side of the mountains, surface winds bring moisture onshore where the rising terrain induces a second extensive rainfall. This pattern usually lingers for a few days before the low weakens and the upper westerlies return.

To an eclipse watcher, the most important feature of December weather systems is the traveling high-pressure systems that move across South Africa. When an anticyclone is centered over Mozambique and eastern South Africa, the Botswana low is weakened and pushed well to the north over Angola. Winds are generally easterly to northeasterly across the eclipse track and fine weather dominates. If the pattern persists for an extended period, it can bring extreme dryness and even drought. Even in the wet patterns described above, dry areas can be found in the midst of the highs that characteristically inhabit the east end of the eclipse track and that is where the best weather prospects are to be found.

Discussion of the weather systems affecting South Africa would not be complete without mention of tropical cyclones. December is within the cyclone season; on average, there are eleven per year in the southwest Indian Ocean. The cyclones tend to form northeast of Madagascar and track west and southward, occasionally entering the Mozambique Channel before dying against the African coast or curving back to the east. This is an infrequent occurrence – about once every 2 years–but there are years when two cyclones manage to reach into the Channel. They may even be of some benefit to the eclipse seeker, as a sunny anticyclone is usually found over northern South Africa when a cyclone first moves into the Mozambique Channel.

Cloud

Table 20, a collection of useful climatological statistics, Figure 20, a satellite-based map of mean cloud cover, and Figure 21, a graph of mean cloudiness along the track extracted from Figure 20 together present the cloud cover statistics in considerable detail. Two columns in Table 20 speak best for the assessment of cloud cover. These are the percent of possible sunshine and the sum of clear and scattered cloud. The highest value on the eclipse track for the percent of possible sunshine (the number of sunshine hours divided by the time between sunrise and sunset) lies at Beitbridge (59%), a small community in Zimbabwe on the border with South Africa.

Correspondingly high values for the sum of clear and scattered cloud can be found at Shingwedzi in Kruger National Park. The cloud cover map in Figure 20 and the graph in Figure 21 confirm these results, showing a minimum in mean December cloud cover in the same area. Careful examination of these data shows that the best prospects can be found in an area stretching from Beitbridge in Zimbabwe across Kruger National Park in South Africa to Massingir in Mozambique. Beitbridge is an arid area of Zimbabwe, though prodigious rainfalls have been known to occur when a rare tropical cyclone manages to reach the area.

Most of the cloudiness in the southern African summer is convective. Satellite images show that the cloud is generally organized into large systems, so that one area might be nearly overcast while adjacent parts several hundred kilometers away are completely clear. And though areas of thunderstorms tend to have holes in the cloud cover on a finer scale, these are small, difficult to find, and impossible to predict more than a short time ahead. Convective clouds build through the day as the ground warms, with the result that a cooler morning eclipse is likely to have less cloudiness than an afternoon one. The frequency for clear, scattered, broken and overcast cloud in Table 20 accommodates this diurnal trend–the numbers apply to the morning–but the percent of possible sunshine is probably slightly pessimistic.

Temperature

Temperature is intimately related to sunshine, altitude and moisture and the eclipse track shows evidence of all three influences. Low elevation coastal regions in Mozambique have the warmest daily highs, reaching up to the mid thirties Celsius. On the interior plateau average highs range through the upper 20s to the lower 30s, about five degrees cooler than coastal regions. Over coastal Angola, the daily maximum is suppressed because of the presence of cool water offshore. Nighttime temperatures are about 10 degrees below those of the day, a small difference that reflects the relatively short nights, frequent cloudiness and high moisture content of the atmosphere. In the most promising region, from Beitbridge in Zimbabwe to Kruger National Park in South Africa, eclipse observers can expect daily temperatures to range from the lower 20s in the morning to the lower 30s in the afternoon.

Rainfall

The rainfall statistics in Table 20 match the pattern of sunshine, with much lower amounts and frequencies in the sunniest regions. At Beitbridge and Messina, monthly amounts are less than 60 mm. West of Beitbridge the mean monthly precipitation rises abruptly to over 150 mm, and then increases more slowly to over 233 mm at Huambo in Angola. Along the Angolan coast rainfall decreases abruptly (to 61 mm at Lobito) because of the stabilizing influence of the cold Atlantic water. On the opposite side of the continent, east of Messina, the increase in rainfall is more gradual, averaging 88 mm at Shingwedzi in Kruger National Park and just over 100 mm at Maputo and Panda on the Mozambique lowlands. The rainfall minimum is a caused by the frequent anticyclones that inhabit this area and the higher terrain to the southeast that blocks the flow of moisture from the Indian Ocean. Also noteworthy in the statistics in Table 20 is the very low frequency of rainy days at Beitbridge.

Rainfall in southern Africa is highly variable, and the mean statistics in Table 20 greatly smooth the alternating wet and dry years. At Bulawayo the maximum rainfall reported in one set of climatological compilations shows a minimum rainfall of only 3 mm in one December and 273 in another. The range for Messina is between 4 and 141 mm while that for Maputo ranges from 10 to 244. Eclipse chasers could encounter a year with exceptionally dry conditions and very sunny skies or the exact opposite if the weather is wet. Fortunately, and in spite of the range between maximum and minimum, rainfall patterns are actually less variable over northern South Africa than in most other parts of the country and approximately normal conditions can be expected along the eclipse track in about 80% of Decembers.

Winds

Winds are generally light along the eclipse track, save for occasional gusts from thunderstorms and the rare presence of a weakening cyclone along the coast. Mean wind speeds range from 7 to 15 km/h along the track, with lighter values inland away from the coast. The prevailing direction is easterly over Mozambique and much of northern South Africa and then turns to northeasterly and northerly over Zimbabwe, Botswana and Angola. This northerly change is a result of the mean position of the Botswana low. Gale force winds (greater than 36 km/h) are a coastal phenomenon and occur only once or twice in the average December.

Africa Summary

The multiple signals in the climatological record and the satellite imagery point to sites between Beitbridge in Zimbabwe and Massingir in Mozambique as having the best viewing prospects. The single best location, based on the available data, seems to lie at Beitbridge, a community of 6000 on the north bank of the Limpopo River. Because Beitbridge marks the main border crossing between Zimbabwe and South Africa and is very close to where the center line crosses, viewing sites can be selected in either country. The relatively high sun angle and convective nature of the cloudiness, which will be at a minimum for a morning eclipse, suggest a probability of success around 60 percent.

Australian Weather Prospects

Australia Overview

The same features that control the weather along the eclipse track in Africa also operate in Australia, though the equatorial moisture is less important and the influence of the anticyclones is increased. In December the southern part of Australia is entering the sunniest time of year. For the eclipse observer there is a heavy penalty to be paid for these conditions, for the sun is very low, setting during the latter stages of the eclipse, and the eclipse duration is measured in seconds rather than minutes.

As in Africa, Australia's latitude places the eclipse track between the band of high pressure anticyclones that mark the sub-tropics and the equatorial low pressure belt that demarcates Earth's Intertropical Convergence Zone (ITCZ). In December the mean position of the anticyclones can be found close to its most southerly position, just south of the Australian continent. The descending air in the center of the anticyclones suppresses rain-bearing clouds and the south coast is in the early stages of its summer dry season. Farther inland along the eclipse track, in the desert climates of Queensland, New South Wales, and South Australia, rainfall has no pronounced dry or wet season and is generally light throughout the year.

The dry air in the center of the anticyclones does not extend all of the way to the surface, leaving room for low level moisture to gather. This moisture is typically found below a temperature inversion that inhibits vertical motion and so it is difficult to mix wet lower and dry higher level air and evaporate the cloudiness. Because they have only a small vertical extent, rain is uncommon from such clouds and an area may be cloudy and yet dry at the same time.

This position of the high-pressure belt to the south of the continent allows the tropical moisture from the ITCZ to reach into northern Australia. Typically it is well away from the eclipse track but periodically it moves southward to play a prominent role in the weather over South Australia and southern Queensland. Solar heating is extremely high in northern parts of the continent, and a persistent heat low forms south of Port Hedland in Western Australia in response to the abundant sunshine. This Pilbara low is the Australian counterpart of the Botswana low. Its influence waxes and wanes with the flow of the weather, and occasionally strengthens sufficiently to cast a long low-pressure trough southward across the western continent where it plays a role in the formation of cloud and precipitation.

The Countryside

Ceduna, the first community on the track of the eclipse, lies at the edge of the Nullarbor Plain, a flat and infertile landscape that stretches along the Great Australian Bight. Ceduna is reached by the Eyre Highway, one of the world's loneliest roads, stretching 2400 kilometers from Adelaide to Perth. The town itself is a well-appointed popular resting space for long-distance travelers, offers fishing and whale watching, and for astronomers, the Ceduna Radio Observatory operated by the University of Tasmania. This 30-meter antenna is part of the Australian VLBI array.

Roads from Ceduna lead only a short distance inland and there is little opportunity to move if December 4 proves to be cloudy. The Eyre Highway and most local roads tend to run east-west, leaving very little room to maneuver on an eclipse track only 25 km wide. The only route along the center line is from Ceduna to Maltee, a distance of about 20 kilometers.

Farther northeast the eclipse track crosses Lake Gairdner National Park, a vast salt pan famous for the world land speed records that have been attempted and achieved on its flat surface. Nearby is Lake Hart, another blinding-white salt pan set in a red landscape, but this lake boasts the main launching pad for the Woomera Rocket Range on its northeast side, now no longer in use. Eclipse sites in this area, though convenient to reach by way of Highway 87 from Port Augusta to Alice Springs, will have to contend with the heat of the interior and the tireless Australian flies. The centerline lies very close to Koolymilka, though there seems to be little in the way of community any more at this former launching site.

It should be pointed out that access to the region north of Highway 87 is limited because it is part of the Australian government's Woomera Prohibited Area. Anyone planning eclipse viewing from the region will need detailed maps showing the boundaries of the prohibited area.

A little farther along the track is Roxby Downs, a thoroughly modern uranium-mining town (along with gold, silver and copper) with a turbulent history since its construction in 1986. It's close to the north edge of the track but could provide a convenient base. The surrounding area is more settled than the earlier part of the eclipse track, in large part because of the availability of water from artesian wells. In this area, the eclipsed Sun will hang little more than 5 degrees above the horizon with a duration barely over a half-minute.

Beyond Roxby Downs, the eclipse track skirts the Flinders Ranges. Flinders Ranges National Park, south of the track, is one of the most majestic parks in Australia – an old folded landscape of gorges and craggy red mountains. Local Dreamtime stories tell of the serpent that guards the water holes and formed the Flinders' contours by wriggling north to drink dry the salt lakes.

Continuing northward, the shadow path moves past the Flinders Ranges and into the Sturt Desert. Touching the junction of three states, the track makes a brief passage through New South Wales and then into Queensland. The point where the three states meet is known as Cameron's Corner and the region is the Corner Country. By now, the Sun is very close to the horizon at mid-eclipse, reaching that point just beyond Old Tickalara. The path continues northeastward, right across Australia and into the Pacific Ocean, but gives watchers only a brief view of the eclipsed Sun before it sets.

The Corner Country is a remote part of Australia, well into the Outback. Roads are not paved and are largely deserted. Travel requires food, extra fuel, water and spare parts. The area around Old Tickalara is very dusty; one traveler reports only three homesteads between Cameron's Corner and Thargomindah, a 340-km route that more-or-less follows the eclipse track and its extension northeastward.

LARGE SCALE WEATHER PATTERNS

The position and intensity of the subtropical anticyclones that migrate eastward across the Great Australian Bight control southern Australian weather. The pattern is very similar to that in Africa. As highs approach from the southwest, winds blow first from the southeast and then from the east. At least initially, the southeasterlies are onshore winds and frequently pile low-level clouds against the south shore. As the center of the high reaches the Bight, winds become more easterly and then northeasterly, drawing hot dry air out of the interior and dissipating the coastal clouds.

A low-pressure trough then follows, separating the departing high from the next in the sequence. This trough often contains a cold front, substantially modified after passage over the Bight, but containing enough moisture to create the relatively cloudy climate of coastal regions. Most of the cold front cloudiness does not penetrate far inland before it encounters the dry desert air of the interior and gradually dissipates. This is not an incontrovertible rule however, for the stronger systems are quite capable of pushing cloudy weather right across the length of the eclipse track as far as the Great Dividing Range. Southern Australia has a flat terrain that offers little physical impediment to weather moving in from the south.

The highs and troughs moving alternately past the continent bring a repeating pattern of cloud and sun, cool and hot weather. While the cycle from high to high or trough to trough has a modest degree of regularity (repeating at 5-6 day intervals), the weather in Australia is not a monotonous series of predictable events. Troughs vary in strength, some with fronts, some not. Highs may stall, bringing persistent weather for several weeks before moving on to allow the normal cycle to return. Departures from the average are common and bring weather that is far from average.

In northern Australia, cloud and precipitation are influenced by the Intertropical Convergence Zone that has moved southward onto the continent by December. The presence of the ITCZ signals the beginning of "the Wet," a season of frequent and widespread showers and thundershowers. Moisture from the ITCZ can be induced to flow southward under the right conditions, but typically remains in northern regions away from the eclipse path.

One of the elements that may induce a southward excursion of the moisture is a strengthening in the Pilbara heat low. As the low intensifies, northerly winds increase and push tropical moisture southward across Western Australia. This moisture is typically caught by an inter-anticyclonic trough (with its cold front) and is turned eastward toward the center of the continent and the eclipse track. The combination of tropical moisture and the atmospheric dynamics of the trough can create a large area of broken cloud that may linger over the track for several days. The combined system often contains enough instability for showers and thunderstorms. Satellite photos show characteristic arched bands of cloud with this type of pattern that persist from their first appearance in the Bight until they pass Adelaide several days later.

The low may also be reinforced by a tropical cyclone moving onto the continent from the Indian Ocean, with dramatic consequences for cloud cover across much of the country. This is a relatively uncommon event, occurring only about once a year. On December 6, 2000, Tropical Cyclone Sam moved into northwestern Australia, joined the Pilbara low, and pushed a large area of cloud southward for the next five days. In what might be a good omen for this type of meteorological event, the cloud did not reach the south coast and only the northern part of the eclipse track was affected.

CLOUD

Figure 20 shows the regional distribution of mean daytime cloudiness along the eclipse track. This map is an average of November and December satellite observations combined to better represent the cloudiness around the eclipse date in early December. Because the satellite cloud statistics are derived from fixed algorithms, the results are free of many of the biases that are contained in human observations of cloud cover and form a better comparison of cloud prospects in Australia versus those in Africa. In Australia, coastal regions are the cloudiest and there is a gradual drying along the track inland, at least as far as the Queensland-New South Wales border. The improvement is substantial – there are more than twice as many sunny days in the interior as on the coast. Figure 21 is derived from the satellite data in Figure 20 and graphs the cloud cover along the centerline in more detail.

Figure 22 shows contours of the number of clear days in December based on surface observations of cloud cover from Australian weather stations. The frequency of clear days increases from 8 to 11 days per month at Ceduna where the eclipse track first reaches land, to 19 days at the end of the track where the Sun sets at mid eclipse. Values for the percent of possible sunshine (Table 20) reach almost 80% at Oodnadatta, and nearly 75% at Woomera—statistics 10 to 15 percent better than the best in Africa. While Australia quite clearly has sunnier weather, the low altitude of the Sun will require a nearly clear horizon in order for the eclipse to be seen. This is a much more stringent requirement than a low frequency of cloud cover at midday and the choice of Australia versus Africa will be a tough one to make.

The seasonal frequency of completely clear skies at sunset ranges from 15 to 35 percent (Table 20) with the larger amount inland. Scattered cloudiness (up to half of the sky covered) will bring a modest probability that the horizon will be obscured. Some of this cloud will be convective and will dissipate as the Sun approaches the horizon and temperatures fall, but this cannot be counted on if a cold front has made a recent passage through the area. Larger cloud amounts, when more than half the sky is covered, will bring progressively slimmer chances of seeing the eclipse. If we make the reasonable assumption that there is a one-third chance that scattered cloud will block the Sun and a two-thirds chance for broken cloud, then the chances of seeing the eclipse will range from about 55% on the coast to about 65% inland. All in all, the prospects for a successful horizon eclipse in Australia likely exceed the best chances in Africa by about five to eight percent. To achieve this advantage, a site inland from the coast must be selected.

TEMPERATURE

Temperatures in the interior of Australia can become very hot – similar to those encountered in southern Turkey during the 1999 eclipse. Daytime highs (Table 20) average in the low to mid 30's Celsius (90's F), and records climb into the mid and upper 40's Celsius (100's F). Cooler readings prevail along the coast where sea breezes cut off the daytime heating and bring pleasant relief.

Departing anticyclones bring hot northerly winds and coastal regions can see temperatures nearly as great as the interior. In due course, another high-pressure system will approach from the west and winds will turn to the south. This brings temperate ocean air inland, resulting in an abrupt drop in temperature known as a "cool change." A 10° Celsius change in 30 minutes is typical and 20° Celsius is possible.

Though the eclipse occurs while temperatures are falling after the daytime high, equipment set-up will take place at the hottest time of day. Temperatures in the forties occur two or three times on average each month at inland sites. Readings this high require great caution, for heat stroke is a very real possibility. Plenty of water, misters, hats, a slow pace and occasional refuge in shade would be well advised, along with emergency plans if the heat should prove overpowering. Equipment also suffers in such heat and planning should include provision for cameras and telescopes. It is worth noting that official temperature measurements are made in the shade and readings in full sunlight are much higher.

Rainfall

Along the eclipse track, December rainfall is much lighter in Australia than in Africa. For the most part, monthly accumulations range from 10 to 20 mm. Table 20 shows larger amounts farther east along the extension of the eclipse track (where a short total eclipse may only be seen in the seconds before sunset). This precipitation is found along the Great Dividing Range where terrain and the influence of the Pacific Ocean hold sway. Thundershowers are uncommon, occurring only one or two days of the month.

Winds

Winds favor the southerly quadrant along the eclipse path, especially along the coast where a daily sea breeze is common. At Ceduna, 81% of the 6 PM winds are over 20 knots (37 km/h). Inland at Woomera the frequency declines to 46%. These are substantial winds and will demand some accommodation by the eclipse observer, either to seek a little protection or to anchor equipment to reduce vibrations. At the least, observers with sensitive equipment should seek a site at least a few kilometers inland (and 20 to 40 would be better) to take advantage of the decrease in windiness away from the open coast.

The daily sea breezes have long been recognized as a distinctive feature of the local climatology and are given popular local names such as *Eucla Doctor*, *Albany Doctor* or *Fremantle Doctor*. Sea breezes arise because of the temperature difference between the land and the water; the hot air inland rises, drawing air from the cooler South Australian Bight onshore. Sea breezes typically begin in mid-morning along the coast, but the onset can be delayed or even prevented if opposing winds driven by a larger weather system hold them at bay. Depending on the time of onset and the help or hindrance of prevailing winds, sea breezes may extend as far as two or three hundred kilometers inland where they bring a pleasing drop in temperature and an increase in humidity. The effect of the eclipse on these winds will be an interesting observation, as will their effect on shadow bands.

Dust is an unpleasant component of any strong wind in Australia as the dry interior provides a ready supply of material. The initial gust that signals the arrival of the sea breeze is often very strong and then dies quickly to a steadier and lighter flow. Under such circumstances, it may be sufficient to protect equipment for only a short period, but a lingering dust haze may impede the view to the setting sun.

Dust devils, a whirling column of dust, are also common in Australia in the dry season and contribute to the general haziness of the atmosphere. Some can become quite strong and lift a considerable quantity of dirt into the air. Dust storms are also legendary in the country, being driven by particularly strong cold fronts or by the outflow from strong thunderstorms.

Australia Summary

The best weather prospects for viewing the eclipse will be found inland. Cloud cover is at its lowest from the vicinity of Woomera eastward to the end of the track. The higher solar elevation on the coast is not likely to compensate for the greater amount of cloud, leaving the adventurer with a hot and dusty, but ultimately sunnier inland location for best circumstances. Temperatures could be very high and suitable precautions should be taken to protect from heat stroke.

Save for the fact that the eclipse occurs at sunset, Australia would be an easy choice over Africa if weather were the only factor to be considered when selecting a viewing site.

Eclipse Viewing on the Water

Ships and boats offer an opportunity for considerable mobility when seeking those few minutes of sunshine during the critical moments of an eclipse and can often overcome the limitations of a climatologically challenged land-based site. Figure 21, a graph of mean cloudiness along the track, shows that sites off both continents are cloudier than on the land.

Mean cloud cover in the Indian Ocean off Africa rises slowly, but that in the Great Australian Bight is much greater than inland Australia. While ships off either coast would offer the benefits of mobility, a water-based location in the Bight can also bring the advantages that come with a much higher solar altitude.

The Great Australian Bight is a cloudy place in December, though the closer one moves toward landfall on the eclipse track, the better the weather prospects become (Figure 21). Unfortunately the trade-off is steep, with the Sun declining rapidly in the sky as better weather prospects are approached. On the centerline south of Perth the solar elevation will be about 24 degrees while offshore from Ceduna, the Sun will set before fourth contact. The best location along the track can only be determined on eclipse day according to the weather at hand, and so ship-board expeditions should position themselves on the center line in the middle of the Bight in order to be prepared to move either east or west in search of a break in the clouds. Considerable movement may be necessary in order to find such a break, for frontal systems typically stretch across several hundred kilometers and may require many hours of sailing to break free of cloud. A serious attempt cannot be made without access to satellite imagery and a detailed forecast.

Examination of satellite imagery for 1999 and 2000 suggests that the shipboard chances for a view of the Sun on the Australian Bight are slightly lower or equal to that at an inland site in Australia.

Wave heights off Africa average 1.5 to 2 meters and 2 to 2.5 meters off Australia. The values in waters near Africa are similar to those in the Caribbean in 1998. Wave heights are more variable in the Great Australian Bight, reflecting the more stormy nature of weather systems in the region.

WEATHER WEB SITES

World
1. http://www.tvweather.com - Links to current and past weather around the world.
2. http://www.accuweather.com - Forecasts for many cities worldwide.
3. http://www.weatherunderground.com - Good coverage of Africa and Australia.
4. http://www.worldclimate.com/climate/index.htm - World data base of temperature and rainfall.

Africa
1. http://members.nbci.com/rain00/weatab9.html - A site that provides forecast maps for Africa, including moisture and precipitation forecasts up to 6 days in advance. Some expertise in meteorology would be helpful in interpreting the maps and judging the reliability of the predictions.
2. http://www.nottingham.ac.uk/meteosat - Nottingham University site for obtaining satellite imagery over Africa.
3. http://www.sat.dundee.ac.uk/pdus/ - Another site for satellite imagery for Africa, this time from Dundee University. Select the AV subdirectory for visual images and the AI subdirectory for infrared pictures. Images for other places in the world are also available.
4. http://weather.iafrica.com.na/ - The Namibian Meteorological Service, but it seems to point mostly to the South African Weather Service.
5. http://www.zamnet.zm/weather.html - A site for forecasts from the Zambian Meteorological Department.
6. http://weather.utande.co.zw/dbase/zim-weather.idc - The Zimbabwean Weather Service. Short and medium range forecasts for major centres, including Bulawayo.
7. http://www.weathersa.co.za/ - The South African Weather Service, a good place to get maps, forecasts (out to 14 days!), satellite images (pointed to Nottingham) and many other products.

Australia
1. http://www.bom.gov.au - An excellent web site maintained by the Australian Bureau of Meteorology. Satellite images, climatology, surface pressure analyses, and sunshine maps are all available. This is the site to monitor on eclipse day if an Internet connection is available. Note that the Japanese geostationary satellite that provides images over Australia has recently been configured to transmit fewer pictures in order to extend its lifetime a few more years.

OBSERVING THE ECLIPSE

EYE SAFETY AND SOLAR ECLIPSES

B. Ralph Chou, MSc, OD
Associate Professor, School of Optometry, University of Waterloo
Waterloo, Ontario, Canada N2L 3G1

A total solar eclipse is probably the most spectacular astronomical event that most people will ever experience. There is great interest in eclipses, and thousands of astronomers (both amateur and professional) and eclipse enthusiasts travel around the world to observe and photograph them.

A solar eclipse offers students a unique opportunity to see a natural phenomenon that illustrates the basic principles of mathematics and science that are taught through elementary and secondary school. Indeed, many scientists (including astronomers!) have been inspired to study science as a result of seeing a total solar eclipse. Teachers can use eclipses to show how the laws of motion and the mathematics of orbital motion can predict the occurrence of eclipses. The use of pinhole cameras and telescopes or binoculars to observe an eclipse leads to an understanding of the optics of these devices. The rise and fall of environmental light levels during an eclipse illustrate the principles of radiometry and photometry, while biology classes can observe the associated behavior of plants and animals. It is also an opportunity for children of school age to contribute actively to scientific research—observations of contact timings at different locations along the eclipse path are useful in refining our knowledge of the orbital motions of the Moon and Earth, and sketches and photographs of the solar corona can be used to build a three-dimensional picture of the Sun's extended atmosphere during the eclipse.

However, observing the Sun can be dangerous if you do not take the proper precautions. The solar radiation that reaches the surface of the Earth includes ultraviolet (UV) radiation at wavelengths longer than 290 nm, to radio waves in the meter range. The tissues in the eye transmit a substantial part of the radiation between 380 and 1400 nm to the light-sensitive retina at the back of the eye. While environmental exposure to UV radiation is known to contribute to the accelerated aging of the outer layers of the eye and the development of cataracts, the concern over improper viewing of the Sun during an eclipse is for the development of "eclipse blindness" or retinal burns.

Exposure of the retina to intense visible light causes damage to its light-sensitive rod and cone cells. The light triggers a series of complex chemical reactions within the cells which damages their ability to respond to a visual stimulus, and in extreme cases, can destroy them. The result is a loss of visual function which may be either temporary or permanent, depending on the severity of the damage. When a person looks repeatedly or for a long time at the Sun without proper protection for the eyes, this photochemical retinal damage may be accompanied by a thermal injury—the high level of visible light causes heating that literally cooks the exposed tissue. This thermal injury or photocoagulation destroys the rods and cones, creating a small blind area. The danger to vision is significant because photic retinal injuries occur without any feeling of pain (the retina has no pain receptors), and the visual effects do not occur for at least several hours after the damage is done (Pitts, 1993). Viewing the Sun through binoculars, a telescope or other optical devices without proper protective filters can result in thermal retinal injury because of the high irradiance level due to visible light, as well as near infrared radiation, in the magnified image.

The only time that the Sun can be viewed safely with the naked eye is during a total eclipse, when the Moon completely covers the Sun. *It is never safe to look at the partial phases of any eclipse without the proper equipment and techniques.* Even when 99.9% of the Sun's surface (the photosphere) is obscured during the partial phases of a solar eclipse, the remaining crescent Sun is still intense enough to cause a retinal burn, even though illumination levels are comparable to twilight (Chou, 1981, 1996; Marsh, 1982). Failure to use proper observing methods may result in permanent eye damage or severe visual loss. This can have important adverse effects on career choices and earning potential, since it has been shown that most individuals who sustain eclipse-related eye injuries are children and young adults (Penner and McNair, 1966; Chou and Krailo, 1981).

The same techniques for observing the Sun outside of eclipses are used to view and photograph annular solar eclipses and the partly eclipsed Sun (Sherrod, 1981; Pasachoff 2000; Pasachoff & Covington, 1993; Reynolds & Sweetsir, 1995). The safest and most inexpensive method is by projection. A pinhole or small opening is used to form an image of the Sun on a screen placed about a meter behind the opening. Multiple openings in perfboard, a loosely woven straw hat, or even between interlaced fingers can be used to cast a pattern of solar images on a screen. A similar effect is seen on the ground below a broad-leafed tree: the many "pinholes" formed by overlapping leaves creates hundreds of crescent-shaped images. Binoculars or a small telescope mounted on a tripod can also be used to project a magnified image of the Sun onto a white card. All of these methods can be used to provide a safe view of the partial phases of an eclipse to a group of observers, but care must be taken to ensure that no one looks through the device. The main advantage of the projection methods is that nobody is looking directly at the Sun. The disadvantage of the pinhole method is that the screen must be placed at least a meter behind the opening to get a solar image that is large enough to see easily.

The Sun can only be viewed directly when filters specially designed to protect the eyes are used. Most of these filters have a thin layer of chromium alloy or aluminum deposited on their surfaces that attenuates both visible and near-infrared radiation. A safe solar filter should transmit less than 0.003% (density~4.5)[11] of visible light (380 to 780 nm) and no more than 0.5% (density~2.3) of the near-infrared radiation (780 to 1400 nm). Figure 23 shows transmittance curves for a selection of safe solar filters.

One of the most widely available filters for safe solar viewing is shade number 14 welder's glass, which can be obtained from welding supply outlets. A popular inexpensive alternative is aluminized polyester[12] that has been made specially for solar observation. ("Space blankets" and aluminized polyester used in gardening are NOT suitable for this purpose!) Unlike the welding glass, aluminized polyester can be cut to fit any viewing device, and doesn't break when dropped. It has recently been pointed out that some aluminized polyester filters may have large (up to approximately 1 mm in size) defects in their aluminum coatings that may be hazardous. A microscopic analysis of examples of such defects shows that despite their appearance, the defects arise from a hole in one of the two aluminized polyester films used in the filter. There is no large opening completely devoid of the protective aluminum coating. While this is a quality control problem, the presence of a defect in the aluminum coating does not necessarily imply that the filter is hazardous. When in doubt, an aluminized polyester solar filter that has coating defects larger than 0.2 mm in size, or more than a single defect in any 5 mm circular zone of the filter, should not be used.

An alternative to aluminized polyester solar filter material that has become quite popular is "black polymer" in which carbon particles are suspended in a resin matrix. This material is somewhat stiffer than polyester and requires a special holding cell if it is to be used at the front of binoculars, telephoto lenses or telescopes. Intended mainly as a visual filter, the polymer gives a yellow image of the Sun (aluminized polyester produces a blue-white image). This type of filter may show significant variations in density of the tint across its extent; some areas may appear much lighter than others. Lighter areas of the filter transmit more infrared radiation than may be desirable. A recent development is a filter that consists of aluminum-coated black polymer. Combining the best features of polyester and black polymer, this new material may eventually replace both as the filter of choice in solar eclipse viewers. The transmittance curve of one of these hybrid filters (Polymer Plus™ by Thousand Oaks Optical) is shown in Figure 23. Another material, Baader AstroSolar Safety Film, can be used for both visual and photographic solar observations. It is an ultrathin resin film with excellent optical quality and less scattered light than most polyester filters.

Many experienced solar observers use one or two layers of black-and-white film that has been fully exposed to light and developed to maximum density. The metallic silver contained in the film emulsion is the protective filter; however any black-and-white negative with images in it is not suitable for this purpose. More recently, solar observers have used floppy disks and compact disks (CDs and CD-ROMs) as protective filters by covering the central openings and looking through the disk media. However, the optical quality of the solar image formed by a floppy disk or CD is relatively poor compared to aluminized polyester or welder's glass. Some CDs are made with very thin aluminum coatings which are not safe–if you can see through the CD in normal room lighting, don't use it!! No filter should be used with an optical device (e.g., binoculars, telescope, camera) unless it has been specifically designed for that purpose and is mounted at the front end. Some sources of solar filters are listed below.

Unsafe filters include color film, black-and-white film that contains no silver, film negatives with images on them, smoked glass, sunglasses (single or multiple pairs), photographic neutral density filters and polarizing filters. Most of these transmit high levels of invisible, infrared radiation which can cause a thermal retinal burn (see Figure 23). The fact that the Sun appears dim, or that you feel no discomfort when looking at the Sun through the filter, is no guarantee that your eyes are safe. Solar filters designed to thread into eyepieces that are often provided with inexpensive telescopes are also unsafe. These glass filters often crack unexpectedly from overheating when the telescope is pointed at the Sun, and retinal damage can occur faster than the observer can move the eye from the eyepiece. Avoid unnecessary risks. Your local planetarium, science center, or amateur astronomy club can provide additional information on how to observe the eclipse safely.

---

[11] In addition to the term transmittance (in percent), the energy transmission of a filter can also be described by the term density (unitless) where density 'd' is the common logarithm of the reciprocal of transmittance 't' or $d = \log_{10}[1/t]$. A density of '0' corresponds to a transmittance of 100%; a density of '1' corresponds to a transmittance of 10%; a density of '2' corresponds to a transmittance of 1%, etc....

[12] Aluminized polyester is popularly known to as mylar. DuPont actually owns the trademark "Mylar™" and does not manufacture this material for use as a solar filter.

There are some concerns that UVA radiation (wavelengths between 315 and 380 nm) in sunlight may also adversely affect the retina (Del Priore, 1991). While there is some experimental evidence for this, it only applies to the special case of aphakia, where the natural lens of the eye has been removed because of cataract or injury, and no UV-blocking spectacle, contact or intraocular lens has been fitted. In an intact normal human eye, UVA radiation does not reach the retina because it is absorbed by the crystalline lens. In aphakia, normal environmental exposure to solar UV radiation may indeed cause chronic retinal damage. However, the solar filter materials discussed in this article attenuate solar UV radiation to a level well below the minimum permissible occupational exposure for UVA (ACGIH, 1994), so an aphakic observer is at no additional risk of retinal damage when looking at the Sun through a proper solar filter.

In the days and weeks before a solar eclipse occurs, there are often news stories and announcements in the media, warning about the dangers of looking at the eclipse. Unfortunately, despite the good intentions behind these messages, they frequently contain misinformation, and may be designed to scare people from seeing the eclipse at all. However, this tactic may backfire, particularly when the messages are intended for students. A student who heeds warnings from teachers and other authorities not to view the eclipse because of the danger to vision, and learns later that other students did see it safely, may feel cheated out of the experience. Having now learned that the authority figure was wrong on one occasion, how is this student going to react when other health-related advice about drugs, AIDS, or smoking is given? Misinformation may be just as bad, if not worse than no information (Pasachoff, 2001).

Remember that the *total* phase of an eclipse can and should be seen without any filters, and certainly never by projection! It is completely safe to do so. Even after observing 15 solar eclipses, I find the naked eye view of the totally eclipsed Sun awe-inspiring. I hope you will also enjoy the experience.

SOURCES FOR SOLAR FILTERS

A brief list of sources for solar filters in North America appears below. For additional sources, see advertisements in *Astronomy* and/or *Sky & Telescope* magazines. The inclusion of any source on this list does not imply an endorsement of that source by the authors or NASA.

- American Paper Optics, 3080 Bartlett Corporate Drive, Bartlett, TN 38133.   (800)767-8427
- Celestron International, 2835 Columbia St., Torrance, CA 90503.   (310) 328-9560
- Meade Instruments Corporation, 16542 Millikan Ave., Irvine, CA 92714.   (714) 756-2291
- Orion Telescopes and Binoculars, P.O. Box 1815, Santa Cruz, CA 95061-1815.   (800) 447-1001
- Pocono Mountain Optics, 104 NP 502 Plaza, Moscow, PA 18444.   (717) 842-1500
- Rainbow Symphony, Inc., 6860 Canby Ave., #120, Reseda, CA 91335   (800) 821-5122 *
- Thousand Oaks Optical, Box 5044-289, Thousand Oaks, CA 91359.   (805) 491-3642 *
- Khan Scope Centre, 3243 Dufferin Street, Toronto, Ontario, Canada M6A 2T2   (416) 783-4140
- Perceptor Telescopes TransCanada, Brownsville Junction Plaza, Box 38,
    Schomberg, Ontario, Canada L0G 1T0   (905) 939-2313

* sources for inexpensive hand held solar filters and eclipse glasses

IAU SOLAR ECLIPSE EDUCATION COMMITTEE

In order to ensure that astronomers and public health authorities have access to information on safe viewing practices, the Commission on Education and Development of the International Astronomical Union, the international organization for professional astronomers, set up a Solar Eclipse Education Committee. Under Prof. Jay M. Pasachoff of Williams College, the Committee has assembled information on safe methods of observing the Sun and solar eclipses, eclipse-related eye injuries, and samples of educational materials on solar eclipses (see: *http://www.williams.edu/astronomy/IAU_eclipses*).

For more information, contact Prof. Jay M. Pasachoff, Hopkins Observatory, Williams College, Williamstown, MA 01267, USA (e-mail: jay.m.pasachoff@williams.edu). Information on safe solar filters can be obtained by contacting Dr. B. Ralph Chou (e-mail: bchou@sciborg.uwaterloo.ca).

ECLIPSE PHOTOGRAPHY

The eclipse may be safely photographed provided that the above precautions are followed. Almost any kind of camera with manual controls can be used to capture this rare event. However, a lens with a fairly long focal length is recommended to produce as large an image of the Sun as possible. A standard 50 mm lens yields a minuscule 0.5 mm image, while a 200 mm telephoto or zoom produces a 1.9 mm image. A better choice would be one of the small, compact catadioptic or mirror lenses that have become widely available in the past ten years. The focal length of 500 mm is most common among such mirror lenses and yields a solar image of 4.6 mm. With one solar radius of corona on either side, an eclipse view during totality will cover 9.2 mm. Adding a 2x tele-converter will produce a 1000 mm focal length, which doubles the Sun's size to 9.2 mm. Focal lengths in excess of 1000 mm usually fall within the realm of amateur telescopes. If full disk photography of partial phases on 35 mm format is planned, the focal length of the optics must not exceed 2600 mm. However, since most cameras don't show the full extent of the image in their viewfinders, a more practical limit is about 2000 mm. Longer focal lengths permit photography of only a magnified portion of the Sun's disk. In order to photograph the Sun's corona during totality, the focal length should be no longer than 1500 mm to 1800 mm (for 35 mm equipment). However, a focal length of 1000 mm requires less critical framing and can capture some of the longer coronal streamers. For any particular focal length, the diameter of the Sun's image is approximately equal to the focal length divided by 109 (Table 21).

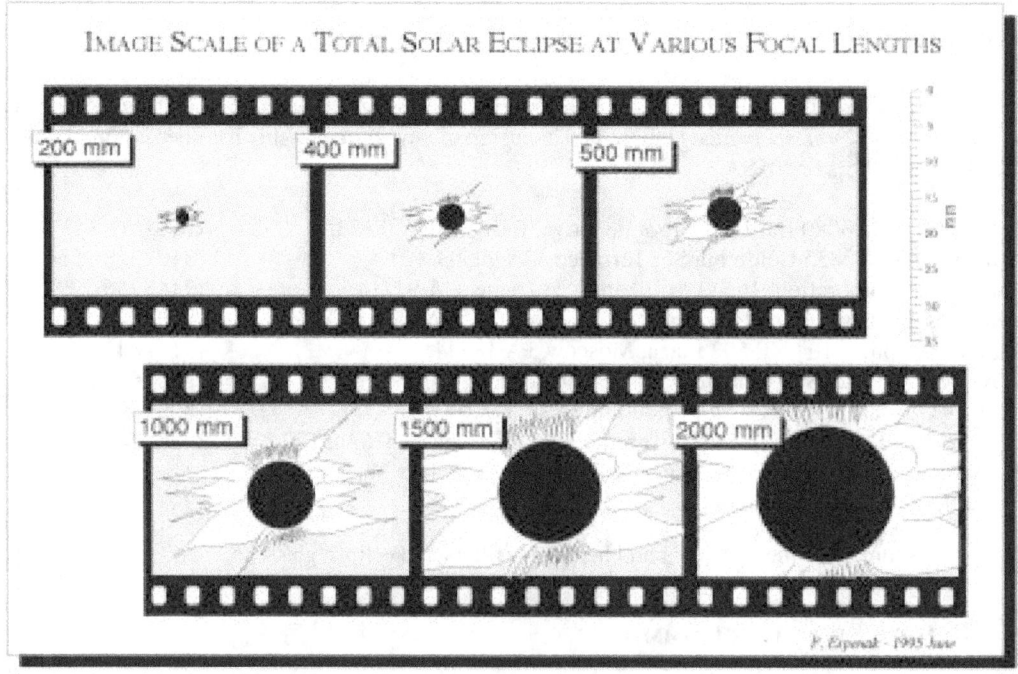

A solar filter must be used on the lens throughout the partial phases for both photography and safe viewing. Such filters are most easily obtained through manufacturers and dealers listed in *Sky & Telescope* and *Astronomy* magazines (see: SOURCES FOR SOLAR FILTERS). These filters typically attenuate the Sun's visible and infrared energy by a factor of 100,000. However, the actual filter factor and choice of ISO film speed will play critical roles in determining the correct photographic exposure. Almost any speed film can be used since the Sun gives off abundant light. The easiest method for determining the correct exposure is accomplished by running a calibration test on the uneclipsed Sun. Shoot a roll of film of the mid-day Sun at a fixed aperture (f/8 to f/16) using every shutter speed between 1/1000 and 1/4 second. After the film is developed, note the best exposures and use them to photograph all the partial phases. The Sun's surface brightness remains constant throughout the eclipse, so no exposure compensation is needed except for the crescent phases which require two more stops due to solar limb darkening. Bracketing by several stops is also necessary if haze or clouds interfere on eclipse day.

Certainly the most spectacular and awe inspiring phase of the eclipse is totality. For a few brief minutes or seconds, the Sun's pearly white corona, red prominences and chromosphere are visible. The great challenge is to obtain a set of photographs which captures some aspect of these fleeting phenomena. The most important point to remember is that during the total phase, all solar filters *must be removed!* The corona has a surface brightness a million times fainter than the photosphere, so photographs of the corona are made without a filter. Furthermore, it is completely safe to view the totally eclipsed Sun directly with the naked eye. No filters are needed and they will only hinder your view. The average brightness of the corona varies inversely with the distance from the Sun's limb. The inner corona is far brighter than the outer corona. Thus, no single exposure can capture its full dynamic range. The best strategy is to choose one aperture or f/number and bracket the exposures over a range of shutter speeds (i.e., 1/1000 down to 1 second). Rehearsing this sequence is highly recommended since great excitement accompanies totality and there is little time to think.

Exposure times for various combinations of film speeds (ISO), apertures (f/number) and solar features (chromosphere, prominences, inner, middle and outer corona) are summarized in Table 22. The table was developed from eclipse photographs made by Espenak as well as from photographs published in *Sky and Telescope*. To use the table, first select the ISO film speed in the upper left column. Next, move to the right to the desired aperture or f/number for the chosen ISO. The shutter speeds in that column may be used as starting points for photographing various features and phenomena tabulated in the 'Subject' column at the far left. For example, to photograph prominences using ISO 400 at f/16, the table recommends an exposure of 1/1000. Alternatively, you can calculate the recommended shutter speed using the 'Q' factors tabulated along with the exposure formula at the bottom of Table 22. Keep in mind that these exposures are based on a clear sky and a corona of average brightness. You should bracket your exposures one or more stops to take into account the actual sky conditions and the variable nature of these phenomena.

An interesting but challenging way to photograph the eclipse is to record its phases all on one frame. This is accomplished by using a stationary camera capable of making multiple exposures (check the camera instruction manual). Since the Sun moves through the sky at the rate of 15 degrees per hour, it slowly drifts through the field of view of any camera equipped with a normal focal length lens (i.e., 35 to 50 mm). If the camera is oriented so that the Sun drifts along the frame's diagonal, it will take over three hours for the Sun to cross the field of a 50 mm lens. The proper camera orientation can be determined through trial and error several days before the eclipse. This will also insure that no trees or buildings obscure the view during the eclipse. The Sun should be positioned along the eastern (left in the northern hemisphere) edge or corner of the viewfinder shortly before the eclipse begins. Exposures are then made throughout the eclipse at ~5-minute intervals. The camera must remain perfectly rigid during this period and may be clamped to a wall or post since tripods are easily bumped. If you're in the path of totality, remove the solar filter during the total phase and take a long exposure (~1 second) in order to record the corona in your sequence. The final photograph will consist of a string of Suns, each showing a different phase of the eclipse.

Finally, an eclipse effect that is easily captured with point-and-shoot or automatic cameras should not be overlooked. Use a kitchen sieve or colander and allow its shadow to fall on a piece of white card-board placed several feet away. The holes in the utensil act like pinhole cameras and each one projects its own image of the Sun. The effect can also be duplicated by forming a small aperture with one's hands and watching the ground below. The pinhole camera effect becomes more prominent with increasing eclipse magnitude. Virtually any camera can be used to photograph the phenomenon, but automatic cameras must have their flashes turned off since this would otherwise obliterate the pinhole images.

Several comments apply to those who choose to photograph the eclipse aboard a cruise ship at sea. Shipboard photography puts certain limits on the focal length and shutter speeds that can be used. It's difficult to make specific recommendations since it depends on the stability of the ship as well as wave heights encountered on eclipse day. Certainly telescopes with focal lengths of 1000 mm or more can be ruled out since their small fields of view would require the ship to remain virtually motionless during totality, and this is rather unlikely even given calm seas. A 500 mm lens might be a safe upper limit in focal length. ISO 400 is a good film speed choice for photography at sea. If it's a calm day, shutter speeds as slow as 1/8 or 1/4 may be tried. Otherwise, stick with a 1/15 or 1/30 and shoot a sequence through 1/1000 second. It might be good insurance to bring a wider 200 mm lens just in case the seas are rougher than expected. As a worst case scenario, Espenak photographed the 1984 total eclipse aboard a 95-foot yacht in seas of 3 feet. He had to hold on with one hand and point his 350 mm lens with the other! Even at that short focal length, it was difficult to keep the Sun in the field. However, any large cruise ship will offer a far more stable platform than this. New image-stabilized lenses from Canon and Nikon may also be helpful aboard ship but allowing the use of slower shutter speeds.

For more information on eclipse photography, observations and eye safety, see FURTHER READING in the BIBLIOGRAPHY.

SKY AT TOTALITY

The total phase of an eclipse is accompanied by the onset of a rapidly darkening sky whose appearance resembles evening twilight about half an hour after sunset. The effect presents an excellent opportunity to view planets and bright stars in the daytime sky. Aside from the sheer novelty of it, such observations are useful in gauging the apparent sky brightness and transparency during totality.

During the total solar eclipse of 2002, the Sun is in southern Ophiuchus. Depending on the geographic location, as many as four naked eye planets and a number of bright stars will be above the horizon within the umbral path. Figure 24 depicts the appearance of the sky during totality as seen from the center line at 06:15 UT. This corresponds to western Zimbabwe near the Botswana border.

The most conspicuous planet visible during totality will be Venus (m_V=–4.5) located 39° west of the Sun in Virgo. Mars (m_V=+1.7) is one and a half degrees west of Venus, but offers a much more difficult target since it will be nearly 500 times fainter. From Zimbabwe, the pair will appear 70° high in the northeast. Compared to Venus, Jupiter (m_V=–2.2) is the next brightest planet but it will be located in the northwestern sky 114° away from the Sun. None of these planets will be visible from Australia since they all set hours before totality begins. However, Mercury (m_V=–0.6) should be discernible from most places along the eclipse track. The innermost planet lies 11° east of the Sun. Since it is nearly at opposition, Saturn (m_V=–0.1) will be below the horizon for most locations along the umbral path.

A number of the bright stars may also be visible during totality. Antares (m_V=+1.06) is 5° south of the Sun while Arcturus (m_V=–0.05) stands 55° to the northwest. Spica (m_V=+0.98) lies 10° west of Venus and Mars. Finally, the great southern stars Alpha (m_V=+0.14) and Beta (m_V=+0.58) Centauri are 45° south of the Sun. Star visibility requires a very dark and cloud free sky during totality.

The following ephemeris [using Bretagnon and Simon, 1986] gives the positions of the naked eye planets during the eclipse. *Delta* is the distance of the planet from Earth (A.U.'s), *App. Mag.* is the apparent visual magnitude of the planet, and *Solar Elong* gives the elongation or angle between the Sun and planet.

Ephemeris: 2002 Dec 04 08:00:00 UT							Equinox = Mean Date
Planet	RA	Declination	Delta	App. Mag.	Apparent Diameter "	Phase	Solar Elong °
Sun	16h41m56s	-22°13'39"	0.98565	-26.7	1947.2	-	-
Moon	16h42m47s	-22°35'01"	0.00244	15.4	1962.8	-0.00	0.4E
Mercury	17h29m23s	-25°14'21"	1.37356	-0.6	4.9	0.95	11.3E
Venus	14h06m16s	-11°02'06"	0.39903	-4.5	41.8	0.24	38.8W
Mars	13h59m38s	-11°16'06"	2.26296	1.7	4.1	0.96	40.2W
Jupiter	09h23m00s	+16°02'32"	4.81623	-2.2	40.9	0.99	113.9W
Saturn	05h45m41s	+22°04'08"	8.07966	-0.1	20.6	1.00	165.2W

For a map of the sky during totality from Australia, see the special web site for the total solar eclipse of 2002: *http://sunearth.gsfc.nasa.gov/eclipse/TSE2002/TSE2002.html*

Contact Timings from the Path Limits

Precise timings of beading phenomena made near the northern and southern limits of the umbral path (i.e., the graze zones), may be useful in determining the diameter of the Sun relative to the Moon at the time of the eclipse. Such measurements are essential to an ongoing project to detect changes in the solar diameter. Due to the conspicuous nature of the eclipse phenomena and their strong dependence on geographical location, scientifically useful observations can be made with relatively modest equipment. A small telescope, short wave radio and portable camcorder are usually used to make such measurements. Time signals are broadcast via short wave stations WWV and CHU, and are recorded simultaneously as the eclipse is videotaped. If a video camera is not available, a tape recorder can be used to record time signals with verbal timings of each event. Inexperienced observers are cautioned to use great care in making such observations. The safest timing technique consists of observing a projection of the Sun rather than directly imaging the solar disk itself. The observer's geodetic coordinates are required and can be measured from USGS or other large scale maps. If a map is unavailable, then a detailed description of the observing site should be included which provides information such as distance and directions of the nearest towns/settlements, nearby landmarks, identifiable buildings and road intersections. The method of contact timing should be described in detail, along with an estimate of the error. The precisional requirements of these observations are ±0.5 seconds in time, 1" (~30 meters) in latitude and longitude, and ±20 meters (~60 feet) in elevation. Commercially available GPS's (Global Positioning Satellite receivers) have comparable positional accuracy as long as the U. S. Department of Defense keeps SA (Selective Availability) turned off. Otherwise, SA degrades the positional accuracy to about ±100. GPS receivers are also a useful source for accurate UT. The International Occultation Timing Association (IOTA) coordinates observers worldwide during each eclipse. For more information, contact:

Dr. David W. Dunham, IOTA
7006 Megan Lane
Greenbelt, MD 20770-3012, USA

Web Site: http://www.lunar-occultations.com/iota
E-mail: dunham@erols.com
Phone: (301) 474-4722

Send reports containing graze observations, eclipse contact and Baily's bead timings, including those made anywhere near or in the path of totality or annularity to:

Dr. Alan D. Fiala
Orbital Mechanics Dept.
U. S. Naval Observatory
3450 Massachusetts Ave., NW
Washington, DC 20392-5420, USA

Plotting the Path on Maps

For high-resolution maps of the umbral path, the coordinates listed in Tables 7 through 10 are conveniently provided in longitude increments of 30' to assist plotting by hand. The coordinates in Table 3 define a line of maximum eclipse at 5-minute increments. If observations are to be made near the limits, then the grazing eclipse zones tabulated in Tables 9 and 10 should be used. A higher resolution table of graze zone coordinates at longitude increments of 7.5' is available via a special 2002 eclipse web site (*http://sunearth.gsfc.nasa.gov/eclipse/TSE2002/TSE2002.html*). Global Navigation Charts (1:5,000,000), Operational Navigation Charts (scale 1:1,000,000) and Tactical Pilotage Charts (1:500,000) of the world are published by the National Imagery and Mapping Agency. Sales and distribution of these maps is through the National Ocean Service (NOS). For specific information about map availability, purchase prices, and ordering instructions, contact the NOS at:

NOAA Distribution Division, N/ACC3
National Ocean Service
Riverdale, MD 20737-1199, USA

phone: 301-436-8301
FAX: 301-436-6829

It is also advisable to check the telephone directory for any map specialty stores in your city/area. They often have large inventories of many maps available for immediate delivery.

IAU Working Group on Eclipses

Professional scientists are asked to send descriptions of their eclipse plans to the Working Group on Eclipses of the International Astronomical Union, so that they can keep a list of observations planned. Send such descriptions, even in preliminary form, to:

International Astronomical Union/Working Group on Eclipses
Prof. Jay M. Pasachoff, Chair   web: www.williams.edu/astronomy/IAU_eclipses
Williams College–Hopkins Observatory   email: jay.m.pasachoff@williams.edu
Williamstown, MA 01267, USA   FAX: (413) 597-3200

The members of the Working Group on Eclipses the Solar Division of the International Astronomical Union are: Jay M. Pasachoff (USA), Chair; F. Clette (Belgium), F. Espenak (USA); Iraida Kim (Russia); Francis Podmore (Zimbabwe); V. Rusin (Slovakia); Jagdev Singh (India); Yoshinori Suematsu (Japan); consultant: J. Anderson (Canada).

Eclipse Data on Internet

NASA Eclipse Bulletins on Internet

To make the NASA solar eclipse bulletins accessible to as large an audience as possible, these publications are also available via the Internet. This was made possible through the efforts and expertise of Dr. Joe Gurman (GSFC/Solar Physics Branch).

NASA eclipse bulletins can be read or downloaded via the World Wide Web using a Web browser (e.g., Netscape, Microsoft Explorer, etc.) from the GSFC SDAC (Solar Data Analysis Center) Eclipse Information home page, or from top-level URL's for the currently available eclipse bulletins themselves:

http://umbra.nascom.nasa.gov/eclipse/   (SDAC Eclipse Information)

http://umbra.nascom.nasa.gov/eclipse/941103/rp.html   (1994 Nov 3)
http://umbra.nascom.nasa.gov/eclipse/951024/rp.html   (1995 Oct 24)
http://umbra.nascom.nasa.gov/eclipse/970309/rp.html   (1997 Mar 9)
http://umbra.nascom.nasa.gov/eclipse/980226/rp.html   (1998 Feb 26)
http://umbra.nascom.nasa.gov/eclipse/990811/rp.html   (1999 Aug 11)
http://umbra.nascom.nasa.gov/eclipse/010621/rp.html   (2001 Jun 21)
http://umbra.nascom.nasa.gov/eclipse/021204/rp.html   (2002 Dec 04)

The original Microsoft Word text files, GIF and PICT figures (Macintosh format) are also available via anonymous ftp. They are stored as BinHex-encoded, StuffIt-compressed Mac folders with .hqx suffixes. For PC's, the text is available in a zip-compressed format in files with the .zip suffix. There are three sub directories for figures (GIF format), maps (JPEG format), and tables (html tables, easily readable as plain text). For example, NASA RP 1383 (Total Solar Eclipse of 1998 February 26 [=980226]) has a directory for these files is as follows:

file://umbra.nascom.nasa.gov/pub/eclipse/980226/RP1383GIFs.hqx
file://umbra.nascom.nasa.gov/pub/eclipse/980226/RP1383PICTs.hqx
file://umbra.nascom.nasa.gov/pub/eclipse/980226/RP1383text.hqx
file://umbra.nascom.nasa.gov/pub/eclipse/980226/RP1383text.zip
file://umbra.nascom.nasa.gov/pub/eclipse/980226/figures   (directory with GIF's)
file://umbra.nascom.nasa.gov/pub/eclipse/980226/maps   (directory with JPEG's)
file://umbra.nascom.nasa.gov/pub/eclipse/980226/tables   (directory with html's)

Other eclipse bulletins have a similar directory format.

Current plans call for making all future NASA eclipse bulletins available over the Internet, at or before publication of each. The primary goal is to make the bulletins available to as large an audience as possible. Thus, some figures or maps may not be at their optimum resolution or format. Comments and suggestions are actively solicited to fix problems and improve on compatibility and formats.

Future Eclipse Paths on Internet

Presently, the NASA eclipse bulletins are published 18 to 36 months before each eclipse. However, there have been a growing number of requests for eclipse path data with an even greater lead time. To accommodate the demand, predictions have been generated for all central solar eclipses from 1991 through 2030. All predictions are based on j=2 ephemerides for the Sun [Newcomb, 1895] and Moon [Brown, 1919, and Eckert, Jones and Clark, 1954]. The value used for the Moon's secular acceleration is n-dot = –26 arc-sec/cy^2, as deduced by Morrison and Ward [1975]. A correction of –0.6" was added to the Moon's ecliptic latitude to account for the difference between the Moon's center of mass and center of figure. The value for ΔT is from direct measurements during the 20th century and extrapolation into the 21st century. The value used for the Moon's mean radius is k=0.272281.

The umbral path characteristics have been predicted at 2-minute intervals of time compared to the 6-minute interval used in *Fifty Year Canon of Solar Eclipses: 1986-2035* [Espenak, 1987]. This should provide enough detail for making preliminary plots of the path on larger scale maps. Global maps using an orthographic projection also present the regions of partial and total (or annular) eclipse. The index page for the path tables and maps is:

http://sunearth.gsfc.nasa.gov/eclipse/SEpath/SEpath.html

Special Web Site for 2002 Solar Eclipse

A special web site is being set up to supplement this bulletin with additional predictions, tables and data for the total solar eclipse of 2002. Some of the data posted there include an expanded version of Tables 9 and 10 (Mapping Coordinates for the Zones of Grazing Eclipse), and local circumstance tables with additional cities as well as for astronomical observatories. Also featured will be higher resolution maps of selected sections of the path of totality and limb profile figures for a range of locations/times along the path. The URL of the special TSE2002 site is:

http://sunearth.gsfc.nasa.gov/eclipse/TSE2002/TSE2002.html

Predictions for Eclipse Experiments

This publication provides comprehensive information on the 2002 total solar eclipse to both the professional and amateur/lay communities. However, certain investigations and eclipse experiments may require additional information which lies beyond the scope of this work. We invite the international professional community to contact us for assistance with any aspect of eclipse prediction including predictions for locations not included in this publication, or for more detailed predictions for a specific location (e.g., lunar limb profile and limb corrected contact times for an observing site).

This service is offered for the 2002 eclipse as well as for previous eclipses in which analysis is still in progress. To discuss your needs and requirements, please contact Fred Espenak (espenak@gsfc.nasa.gov).

Algorithms, Ephemerides and Parameters

Algorithms for the eclipse predictions were developed by Espenak primarily from the *Explanatory Supplement* [1974] with additional algorithms from Meeus, Grosjean and Vanderleen [1966] and Meeus [1982]. The solar and lunar ephemerides were generated from the JPL DE200 and LE200, respectively. All eclipse calculations were made using a value for the Moon's radius of k=0.2722810 for umbral contacts, and k=0.2725076 (adopted IAU value) for penumbral contacts. Center of mass coordinates were used except where noted. Extrapolating from 2001 to 2002, a value for ΔT of 64.7 seconds was used to convert the predictions from Terrestrial Dynamical Time to Universal Time. The international convention of presenting date and time in descending order has been used throughout the bulletin (i.e., *year, month, day, hour, minute, second*).

The primary source for geographic coordinates used in the local circumstances tables is *The New International Atlas* (Rand McNally, 1991). Elevations for major cities were taken from *Climates of the World* (U. S. Dept. of Commerce, 1972).

All eclipse predictions presented in this publication were generated on a Macintosh PowerPC 8500 computer. Word processing and page layout for the publication were done using Microsoft Word v5.1. Figures were annotated with Claris MacDraw Pro 1.5. Meteorological diagrams and tables were prepared using Corel Draw 5.0 and Microsoft Excel 5.0.

The names and spellings of countries, cities and other geopolitical regions are not authoritative, nor do they imply any official recognition in status. Corrections to names, geographic coordinates and elevations are actively solicited in order to update the data base for future eclipses. All calculations, diagrams and opinions presented in this publication are those of the authors and they assume full responsibility for their accuracy.

BIBLIOGRAPHY

REFERENCES

Bretagnon, P., and Simon, J. L., *Planetary Programs and Tables from –4000 to +2800*, Willmann-Bell, Richmond, Virginia, 1986.
Dunham, J. B, Dunham, D. W. and Warren, W. H., *IOTA Observer's Manual*, (draft copy), 1992.
Espenak, F., *Fifty Year Canon of Solar Eclipses: 1986–2035*, NASA RP-1178, Greenbelt, MD, 1987.
Explanatory Supplement to the Astronomical Ephemeris and the American Ephemeris and Nautical Almanac, Her Majesty's Nautical Almanac Office, London, 1974.
Herald, D., "Correcting Predictions of Solar Eclipse Contact Times for the Effects of Lunar Limb Irregularities," *J. Brit. Ast. Assoc.*, 1983, **93**, 6.
Littmann, M., Willcox, K. and Espenak, F. *Totality, Eclipses of the Sun*, Oxford University Press, New York, 1999.
Meeus, J., *Astronomical Formulae for Calculators,* Willmann-Bell, Inc., Richmond, 1982.
Meeus, J., Grosjean, C., and Vanderleen, W., *Canon of Solar Eclipses*, Pergamon Press, New York, 1966.
Morrison, L. V., "Analysis of lunar occultations in the years 1943–1974...," *Astr. J.*, 1979, **75**, 744.
Morrison, L. V., and Appleby, G. M., "Analysis of lunar occultations - III. Systematic corrections to Watts' limb-profiles for the Moon," *Mon. Not. R. Astron. Soc.*, 1981, **196**, 1013.
Pasachoff, J. M. and Nelson, B. O., "Timing of the 1984 Total Solar Eclipse and the Size of the Sun," *Solar Physics*, 1987, **108**, 191-194.
Stephenson, F. R., *Historical Eclipses and Earth's Rotation*, Cambridge/New York: Cambridge University Press, 1997 (p.406).
The New International Atlas, Rand McNally, Chicago/New York/San Francisco, 1991.
van den Bergh, G., *Periodicity and Variation of Solar (and Lunar) Eclipses*, Tjeenk Willink, Haarlem, Netherlands, 1955.
Watts, C. B., "The Marginal Zone of the Moon," *Astron. Papers Amer. Ephem.*, 1963, **17**, 1-951.

METEOROLOGY

Climates of the World, U. S. Dept. of Commerce, Washington DC, 1972.
Griffiths, J.F., ed.,*World Survey of Climatology, vol 10, Climates of Africa*, Elsevier Pub. Co., New York, 1972.
Karoly, David J. and Dayton G. Vincent, (eds.), *Meteorology of the Southern Hemisphere*. American Meteorological Society, Boston, MA, 1998.
International Station Meteorological Climate Summary; Vol 4.0 (CDROM), National Climatic Data Center, Asheville, NC, 1996.
Warren, Stephen G., Carole J. Hahn, Julius London, Robert M. Chervin and Roy L. Jenne, *Global Distribution of Total Cloud Cover and Cloud Type Amounts Over Land*, National Center for Atmospheric Research, Boulder, CO., 1986.
World WeatherDisc (CDROM), WeatherDisc Associates Inc., Seattle, WA, 1990.

Eye Safety

American Conference of Governmental Industrial Hygienists, "Threshold Limit Values for Chemical Substances and Physical Agents and Biological Exposure Indices," ACGIH, Cincinnati, 1996, p.100.
Chou, B. R., "Safe Solar Filters," *Sky & Telescope*, August 1981, 62:2, 119.
Chou, B. R., "Solar Filter Safety," *Sky & Telescope*, February 1998, 95:2, 119.
Chou, B. R., "Eye safety during solar eclipses - myths and realities," in Z. Mouradian & M. Stavinschi (eds.) *Theoretical and Observational Problems Related to Solar Eclipses, Proceedings of a NATO Advanced Research Workshop.* Kluwer Academic Publishers, Dordrecht, 1996, pp. 243-247.
Chou, B. R. and Krailo M. D., "Eye injuries in Canada following the total solar eclipse of 26 February 1979," *Can. J. Optometry*, 1981, 43(1):40.
Del Priore, L. V., "Eye damage from a solar eclipse" in Littmann, M., Willcox, K. and Espenak, F. *Totality, Eclipses of the Sun*, Oxford University Press, New York, 1999, pp. 140-141.
Marsh, J. C. D., "Observing the Sun in Safety," *J. Brit. Ast. Assoc.*, 1982, **92**, 6.
Pasachoff, J. M., "Public Education and Solar Eclipses," in L. Gouguenheim, D. McNally, and J. R. Percy, eds., New Trends in Astronomy Teaching, IAU Colloquium 162 (London), published 1998, Astronomical Society of the Pacific Conference Series, pp. 202-204.
Pasachoff, J. M. "Public Education in Developing Countries on the Occasions of Eclipses," in Astronomy for Developing Countries, IAU special session at the 24th General Assembly, Alan H. Batten, ed., 2001, pp. 101-106.
Penner, R. and McNair, J. N., "Eclipse blindness - Report of an epidemic in the military population of Hawaii," *Am. J. Ophthalmology*, 1966, 61:1452.
Pitts D. G., "Ocular effects of radiant energy," in D. G. Pitts & R. N. Kleinstein (eds.) *Environmental Vision: Interactions of the Eye, Vision and the Environment*, Butterworth-Heinemann, Toronto, 1993, p. 151.

Further Reading

Allen, D., and Allen, C., *Eclipse*, Allen & Unwin, Sydney, 1987.
Astrophotography Basics, Kodak Customer Service Pamphlet P150, Eastman Kodak, Rochester, 1988.
Brewer, B., *Eclipse*, Earth View, Seattle, 1991.
Brunier, S., *Glorious Eclipses*, Cambridge, Nuiv Press, NY, 2001.
Covington, M., *Astrophotography for the Amateur*, Cambridge University Press, Cambridge, 1988.
Espenak, F., "Total Eclipse of the Sun," *Petersen's PhotoGraphic*, June 1991, p. 32.
Golub, L., and Pasachoff, J. M., *The Solar Corona*, Cambridge University Press, Cambridge, 1997.
Golub, L., and Pasachoff, J., *Nearest Star: The Exciting Science of Our Sun*, Harvard University Press, Cambridge, 2001.
Harrington, P. S., *Eclipse!*, John Wiley & Sons, New York, 1997.
Harris, J., and Talcott, R., *Chasing the Shadow,* Kalmbach Pub., Waukesha, 1994.
Irwin, A., *Africa & Madagascar - Total Eclipse 2001 & 2002*, Pradt, Bucks (UK), 2000.
Littmann, M., Willcox, K. and Espenak, F. *Totality, Eclipses of the Sun*, Oxford University Press, New York, 1999.
Mucke, H., and Meeus, J., *Canon of Solar Eclipses: –2003 to +2526*, Astronomisches Büro, Vienna, 1983.
North, G., *Advanced Amateur Astronomy*, Edinburgh University Press, 1991.
Oppolzer, T. R. von, *Canon of Eclipses*, Dover Publications, New York, 1962.
Ottewell, G., *The Under-Standing of Eclipses*, Astronomical Workshop, Greenville, SC, 1991.
Pasachoff, J. M., and Covington, M., *Cambridge Guide to Eclipse Photography*, Cambridge University Press, Cambridge and New York, 1993.
Pasachoff, J. M., "Solar Eclipses and Public Education," International Astronomical Union Colloquium #162: New Trends in Teaching Astronomy, D. McNally, ed., London 1997, in press.
Pasachoff, J. M., *Field Guide to the Stars and Planets*, 4rd edition, Houghton Mifflin, Boston, 2000.
Reynolds, M. D. and Sweetsir, R. A., *Observe Eclipses*, Astronomical League, Washington, DC, 1995.
Sherrod, P. C., *A Complete Manual of Amateur Astronomy*, Prentice-Hall, 1981.
Steel, D., *Eclipse*, Joseph Henery Press, Washington DC, 2001.
Zirker, J. B., *Total Eclipses of the Sun*, Princeton University Press, Princeton, 1995.

Total Solar Eclipse of 2002 Dec 04

Figure 1: Orthographic Projection Map of The Eclipse Path

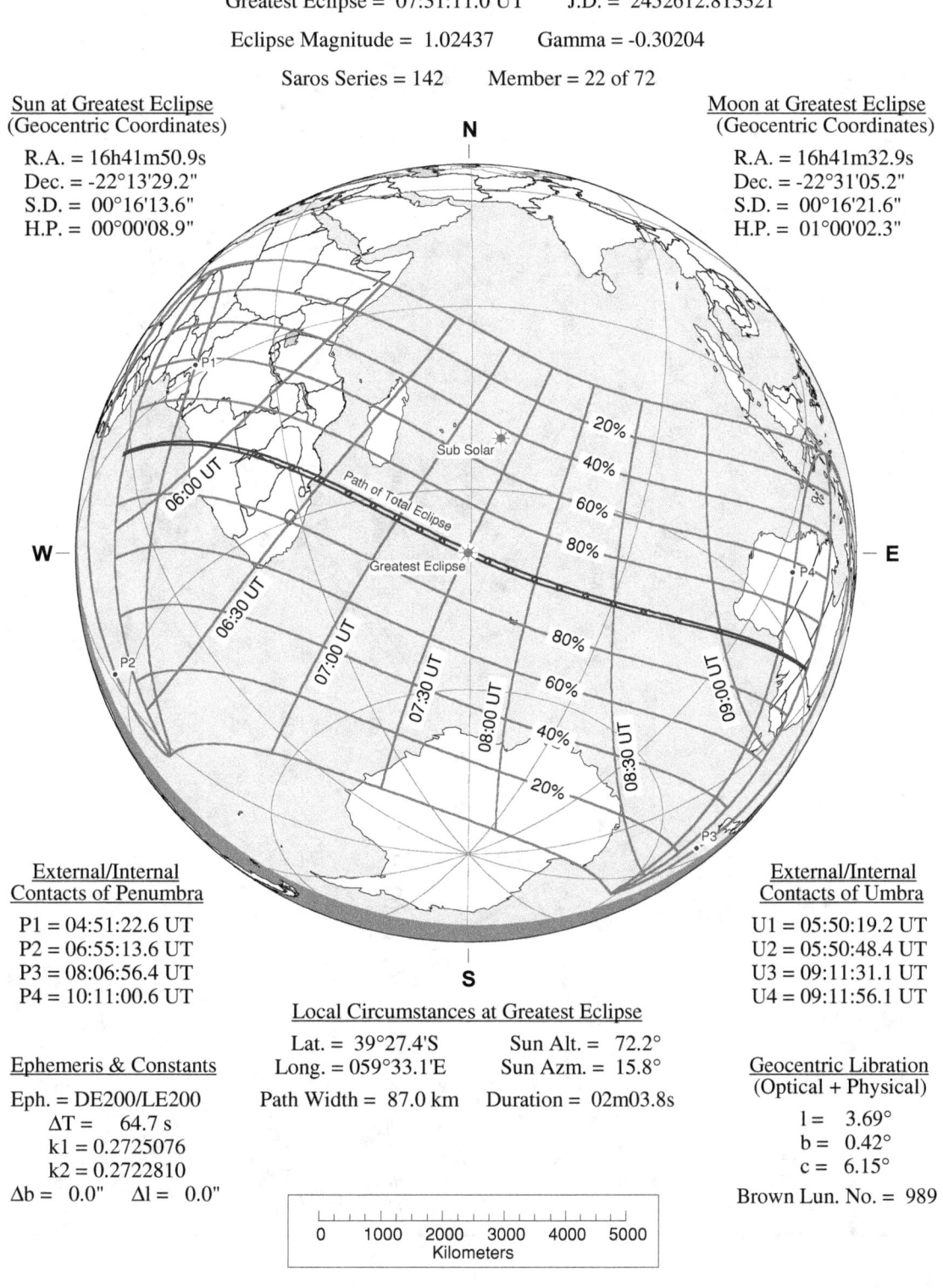

NASA 2002 Eclipse Bulletin (F. Espenak & J. Anderson)

Total Solar Eclipse of 2002 December 04

FIGURE 2: THE ECLIPSE PATH THROUGH AFRICA

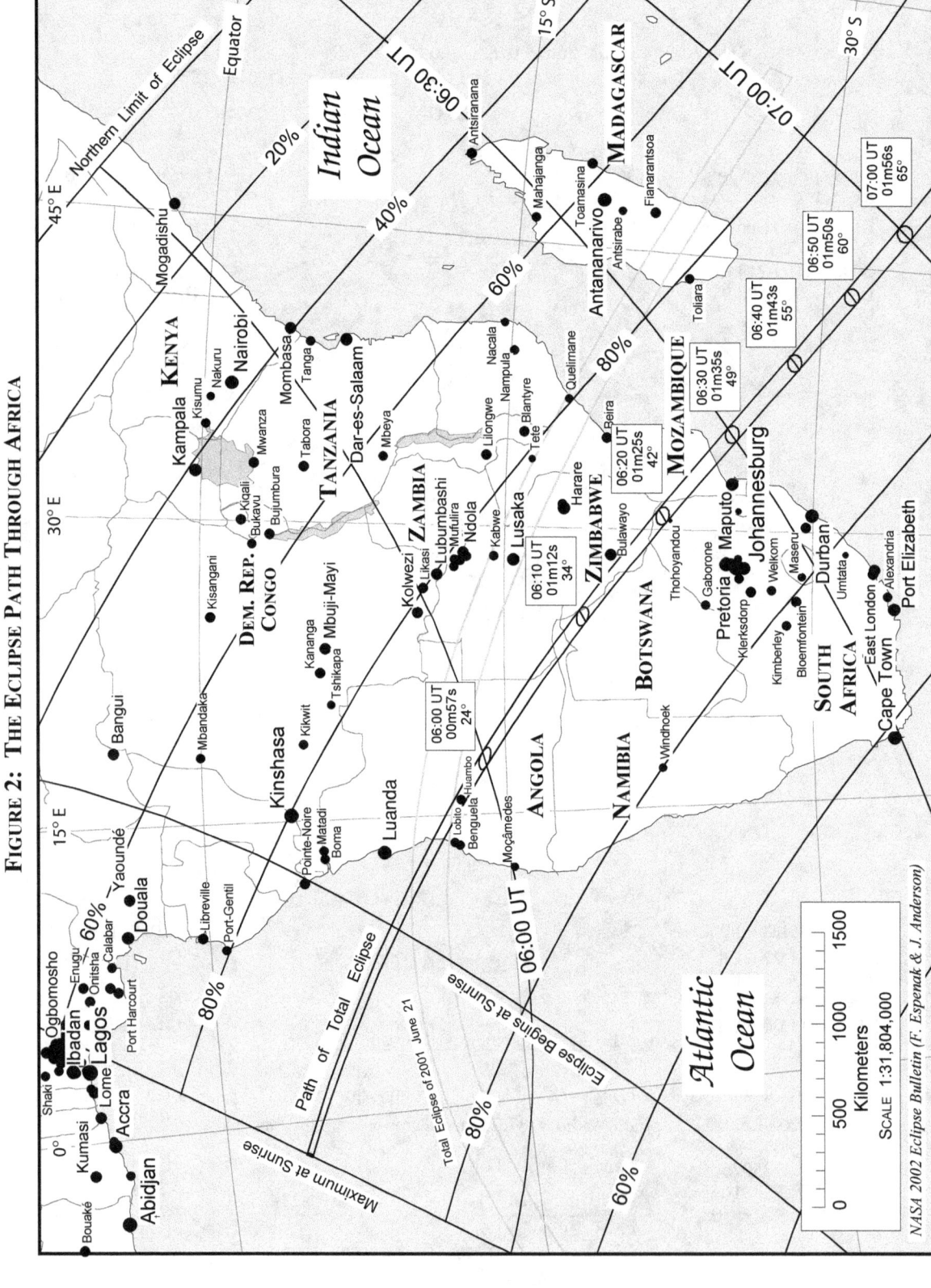

Total Solar Eclipse of 2002 December 04

Figure 3: The Eclipse Path Through Western Africa

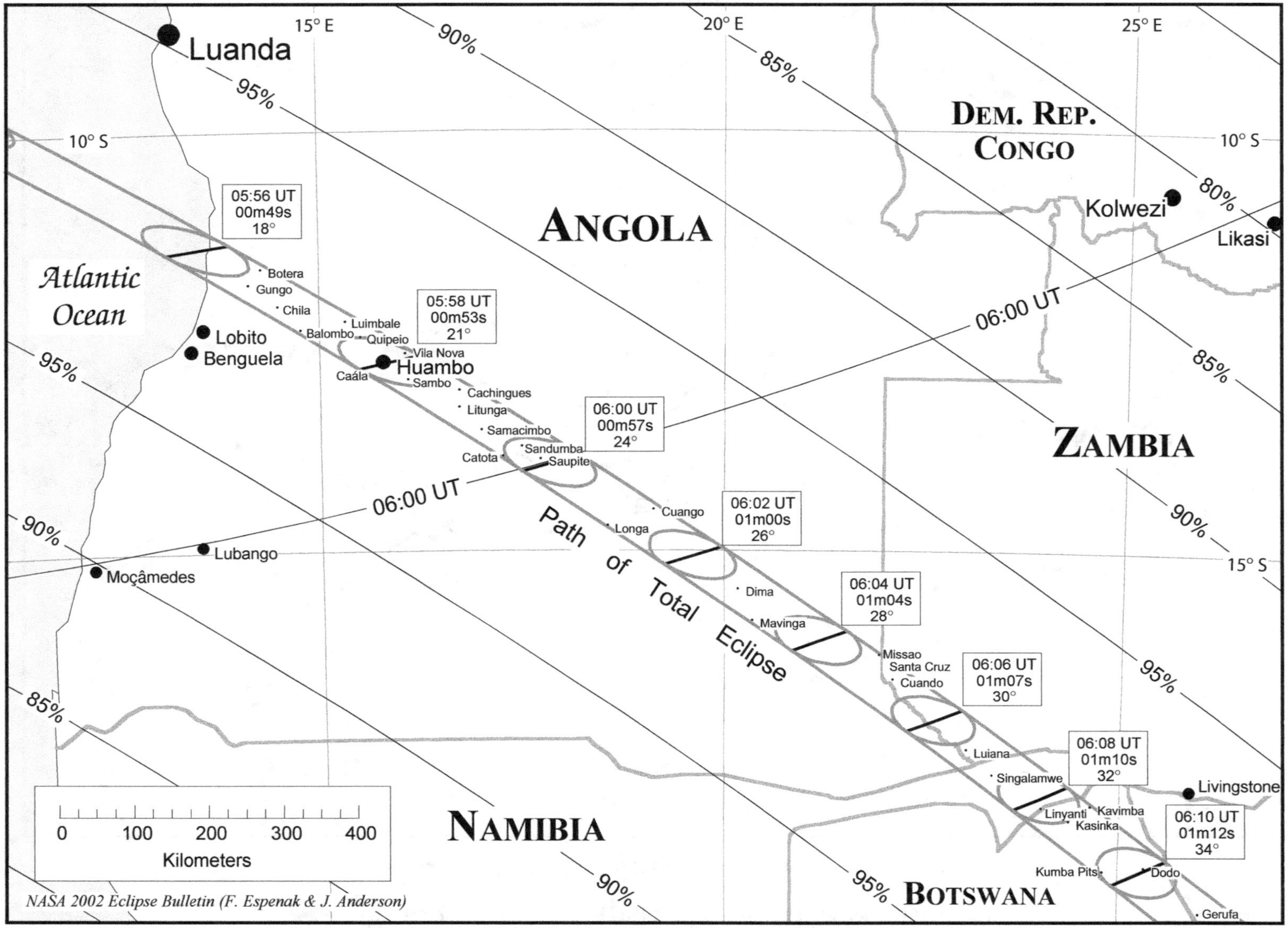

NASA 2002 Eclipse Bulletin (F. Espenak & J. Anderson)

Total Solar Eclipse of 2002 December 04

Figure 4: The Eclipse Path Through Eastern Africa

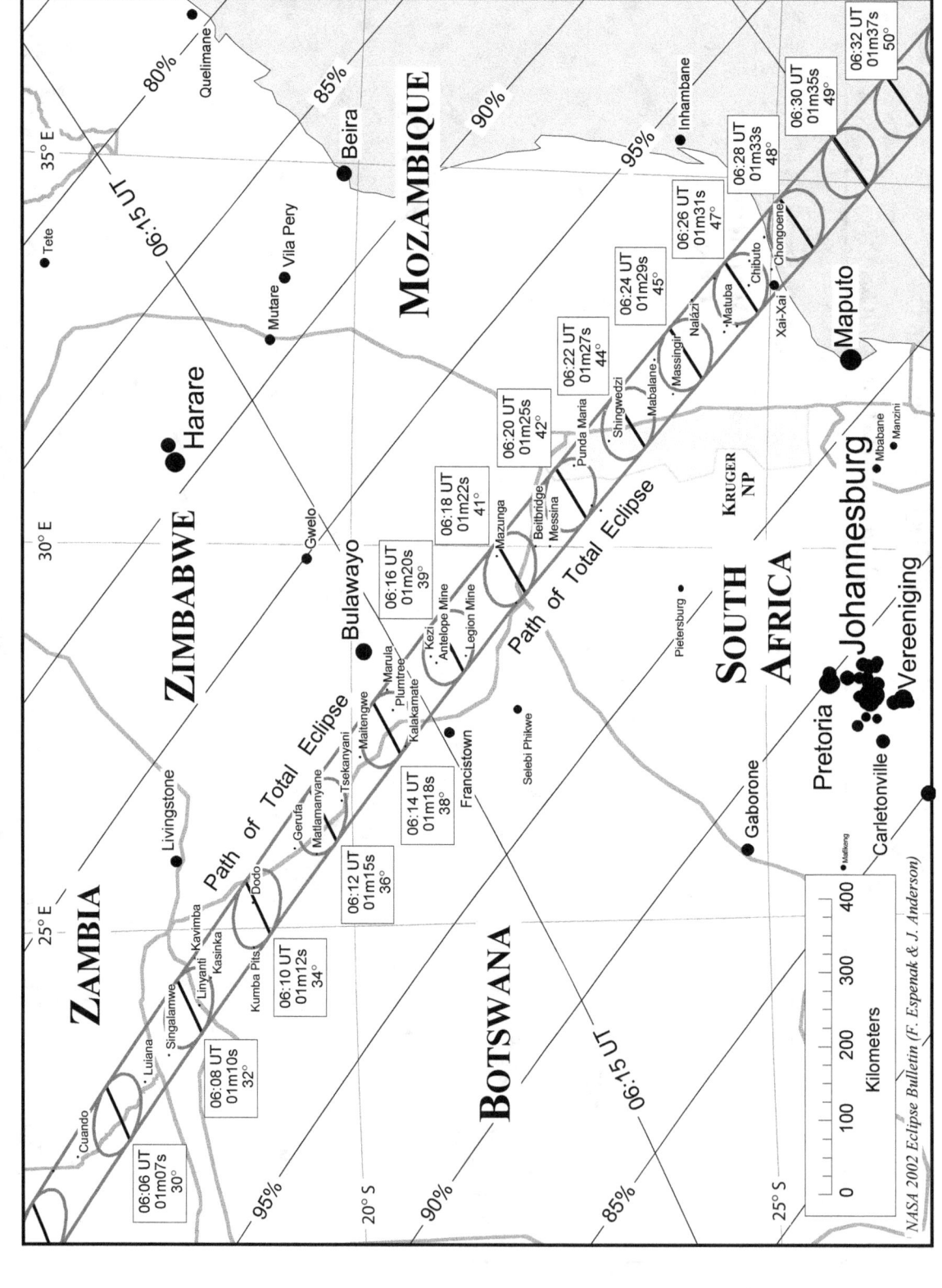

NASA 2002 Eclipse Bulletin (F. Espenak & J. Anderson)

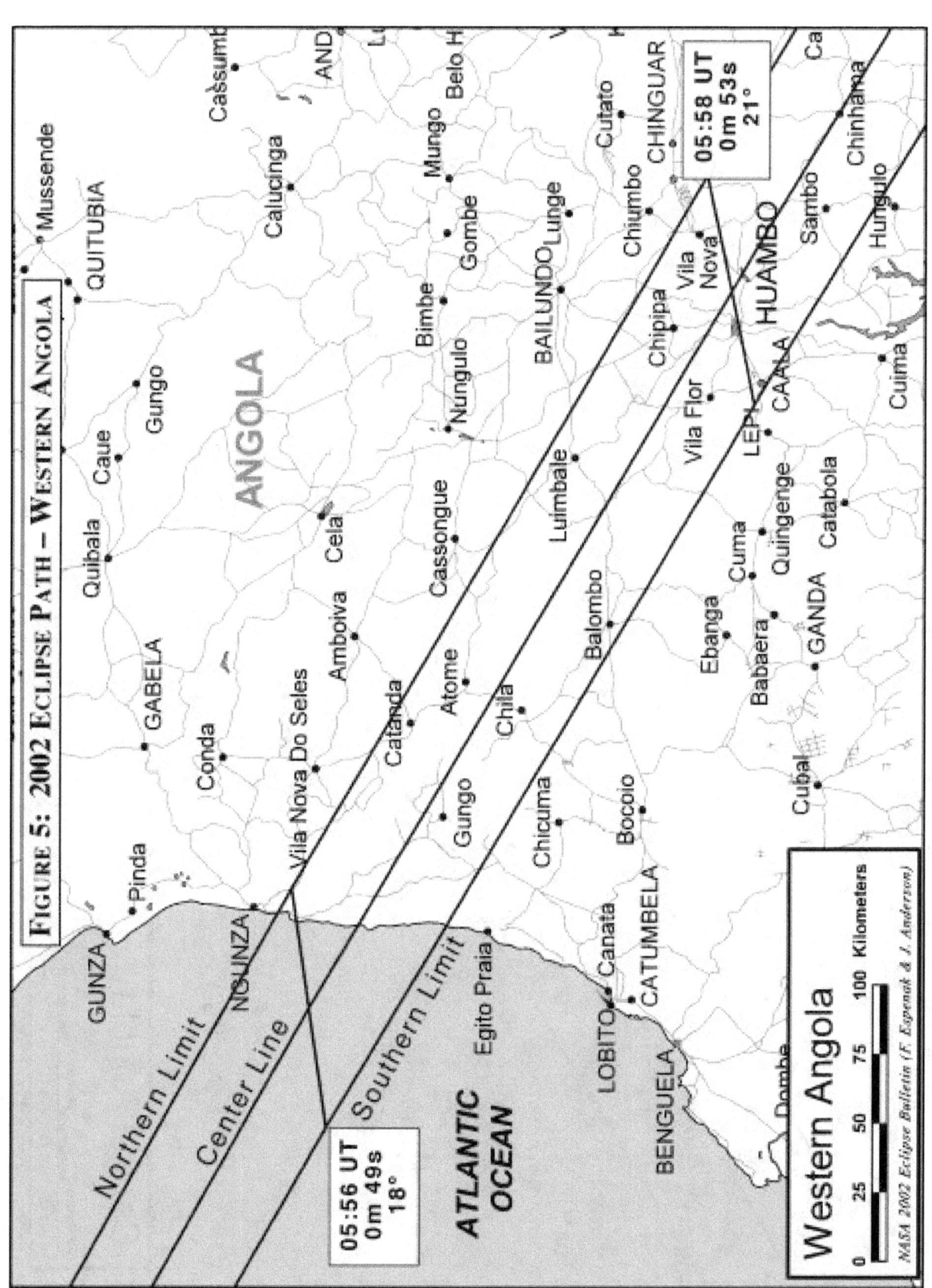

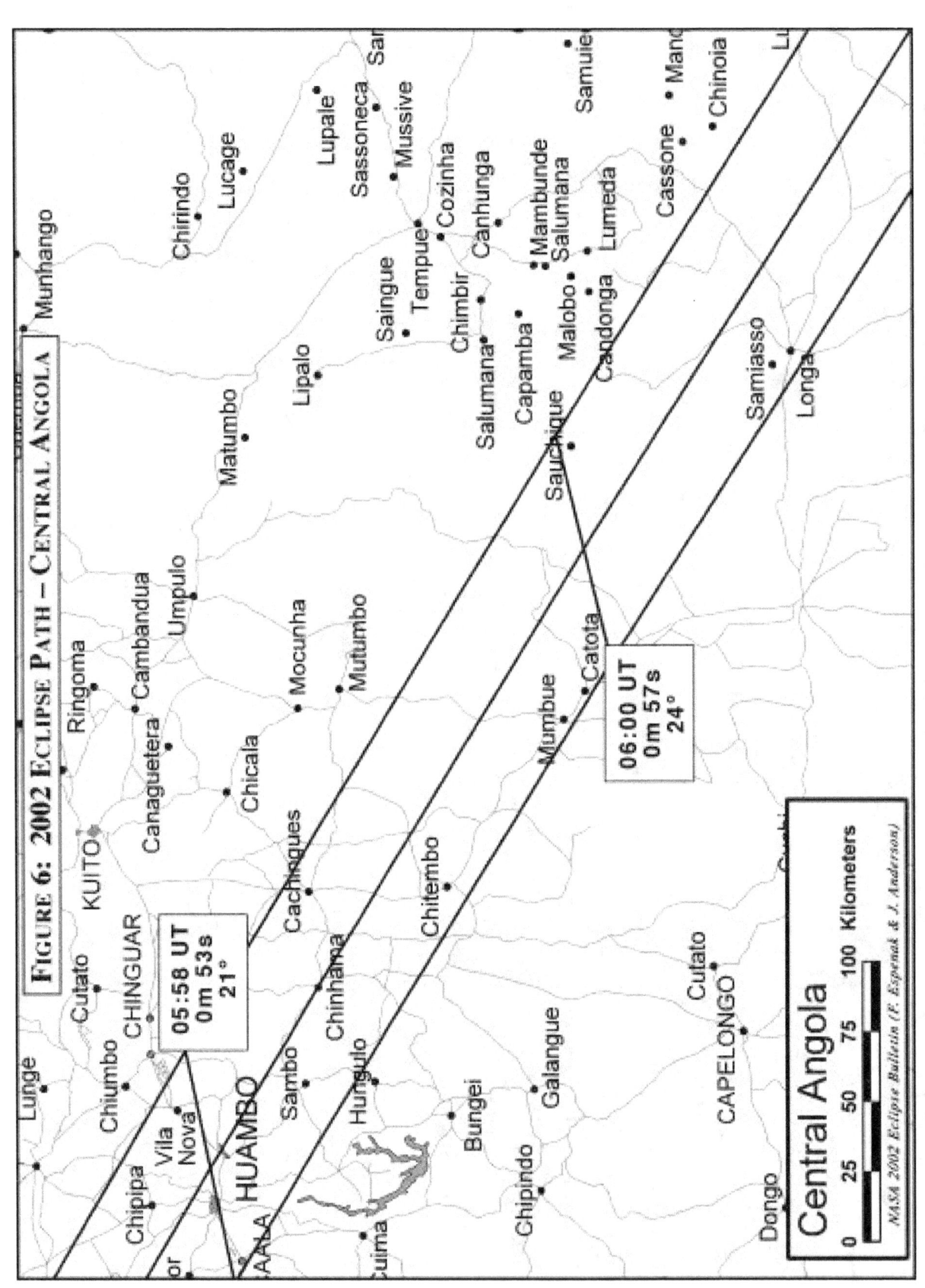

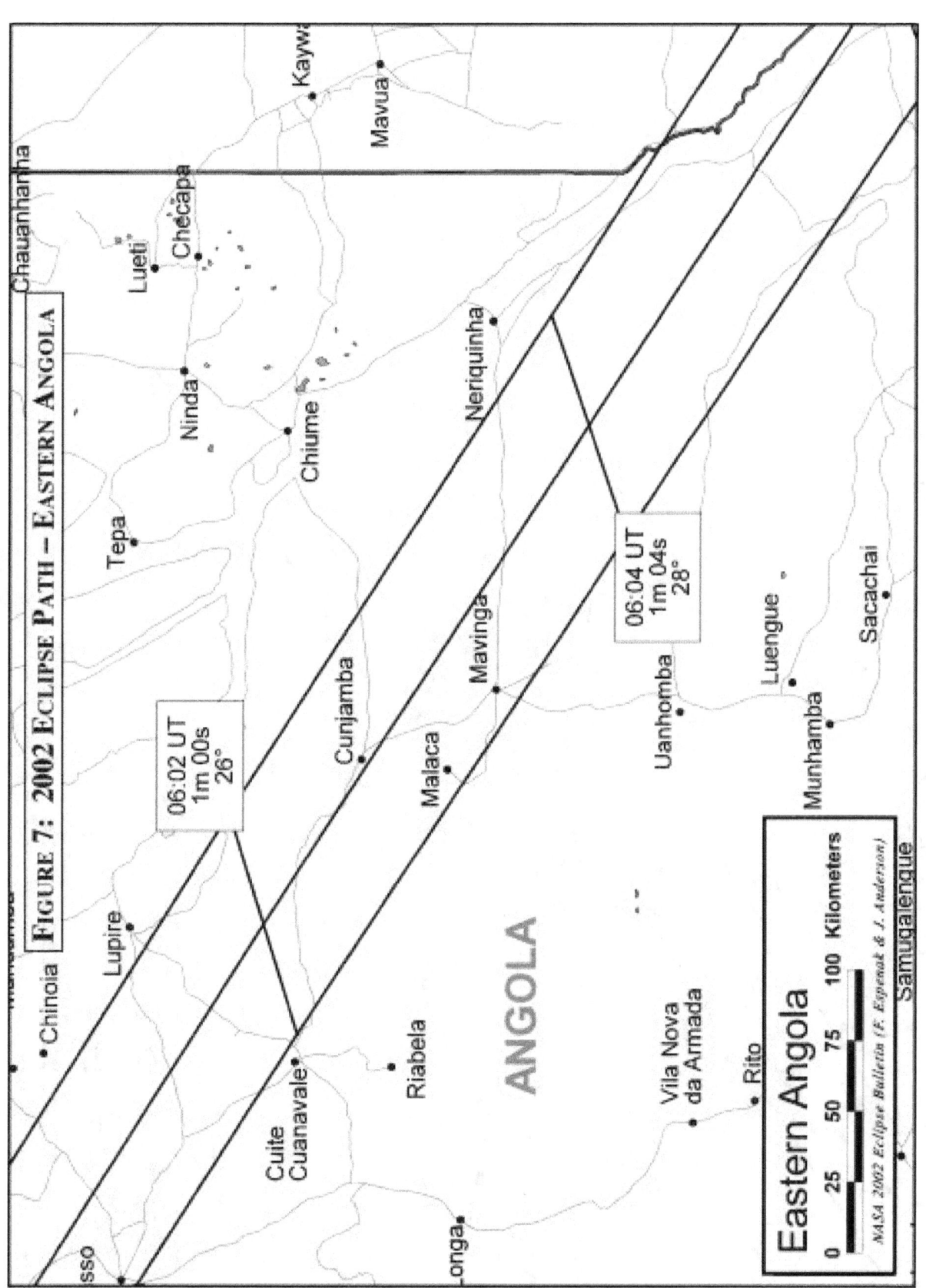

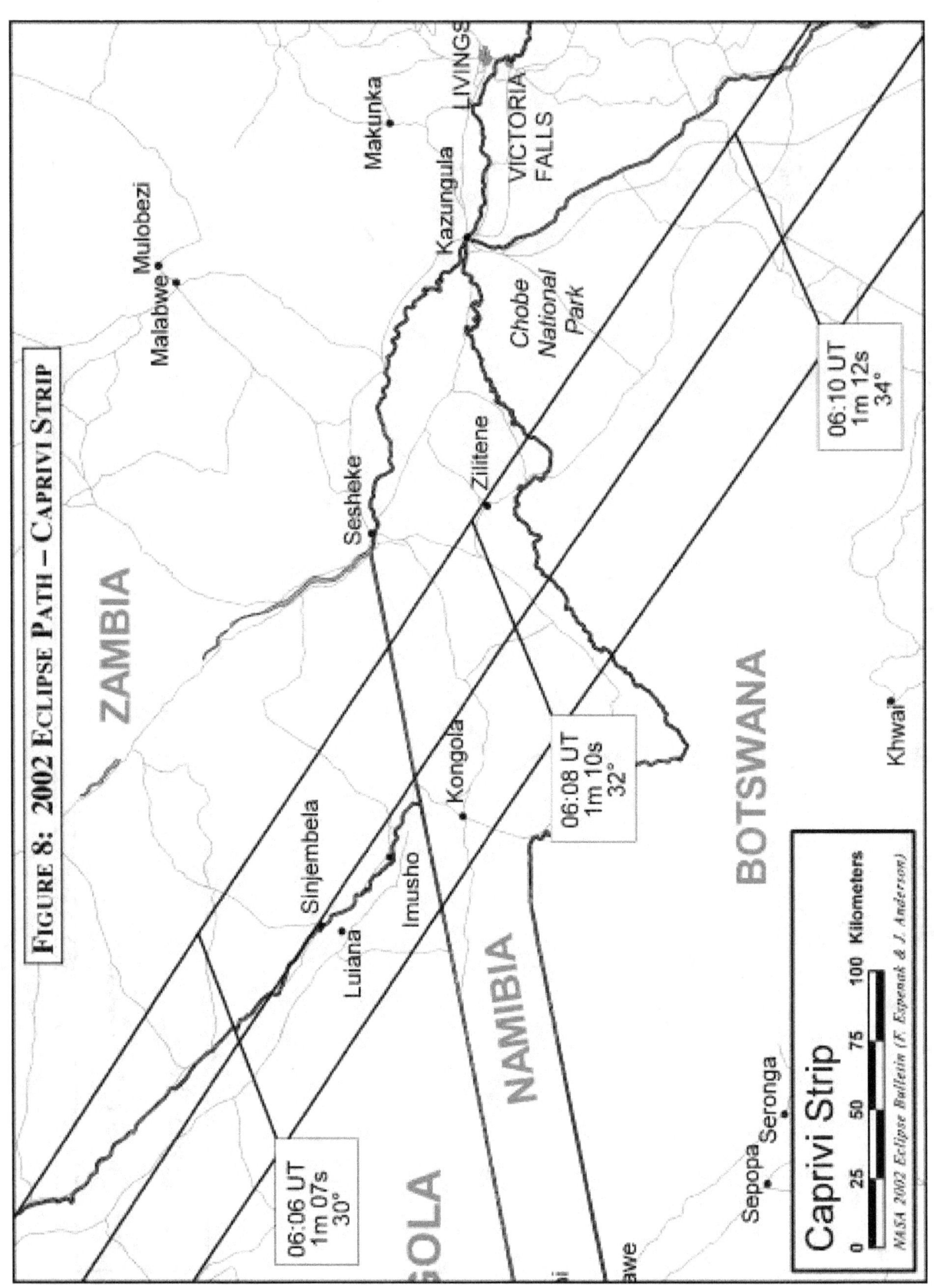

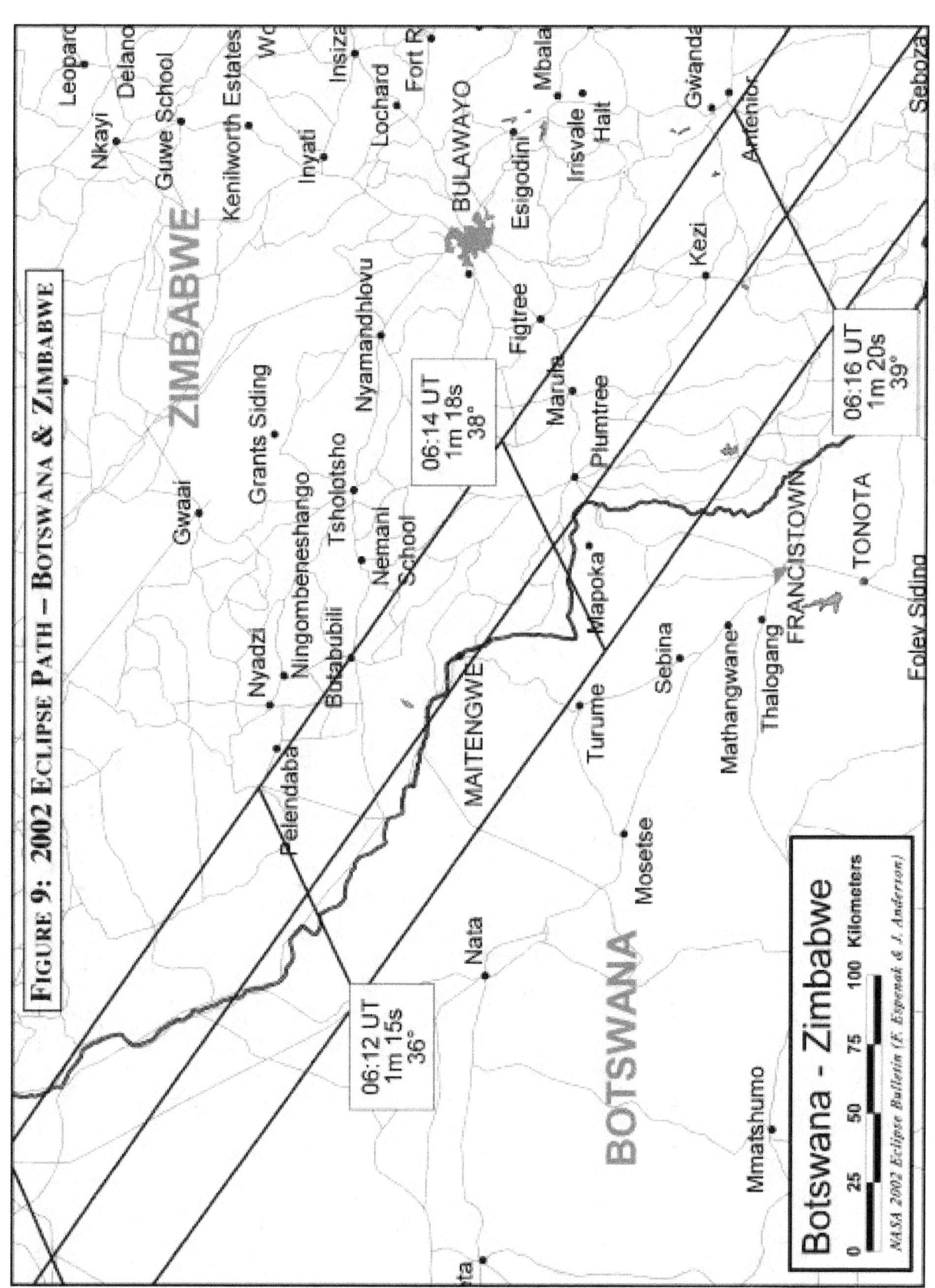

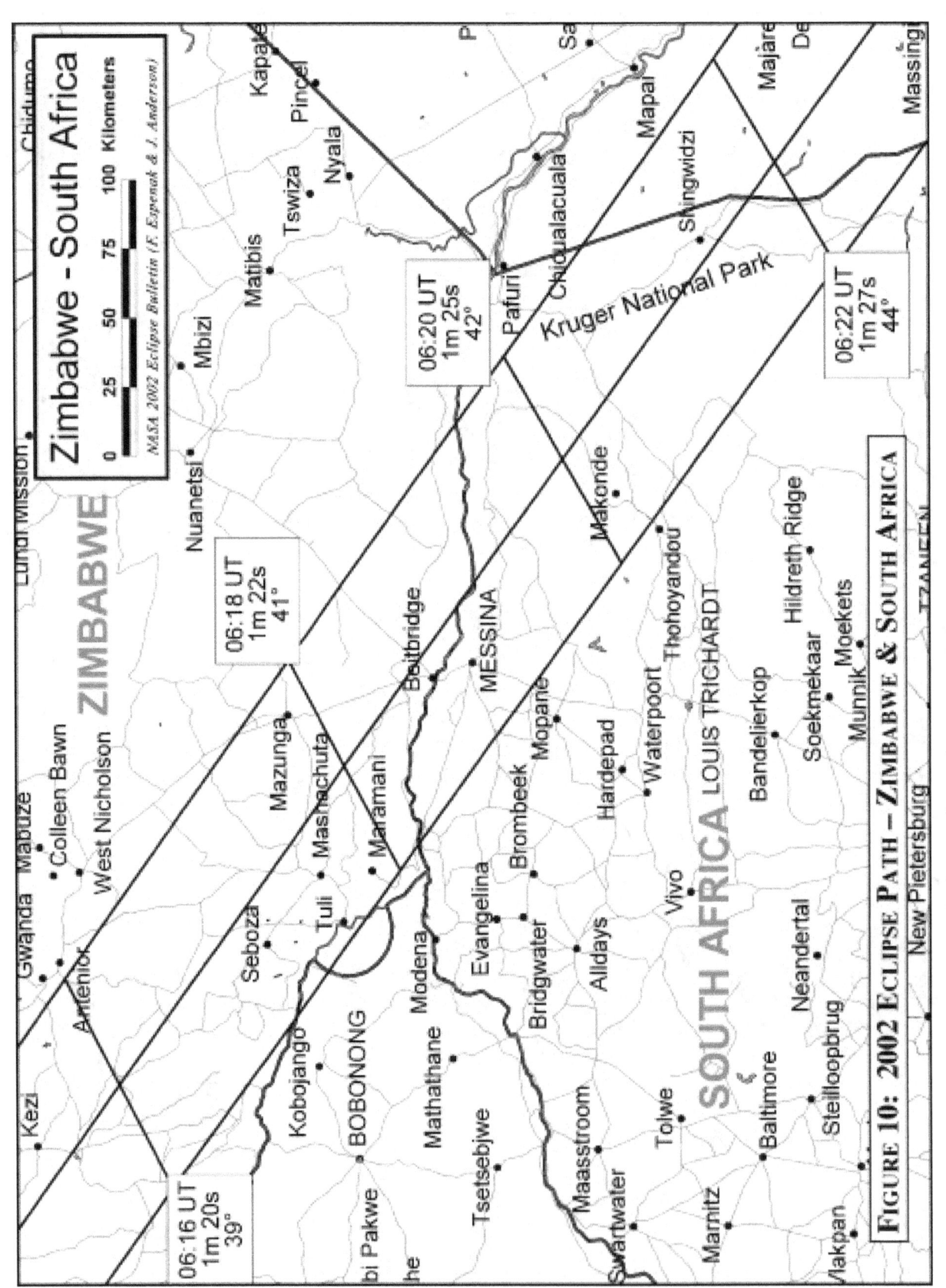

Figure 10: 2002 Eclipse Path – Zimbabwe & South Africa

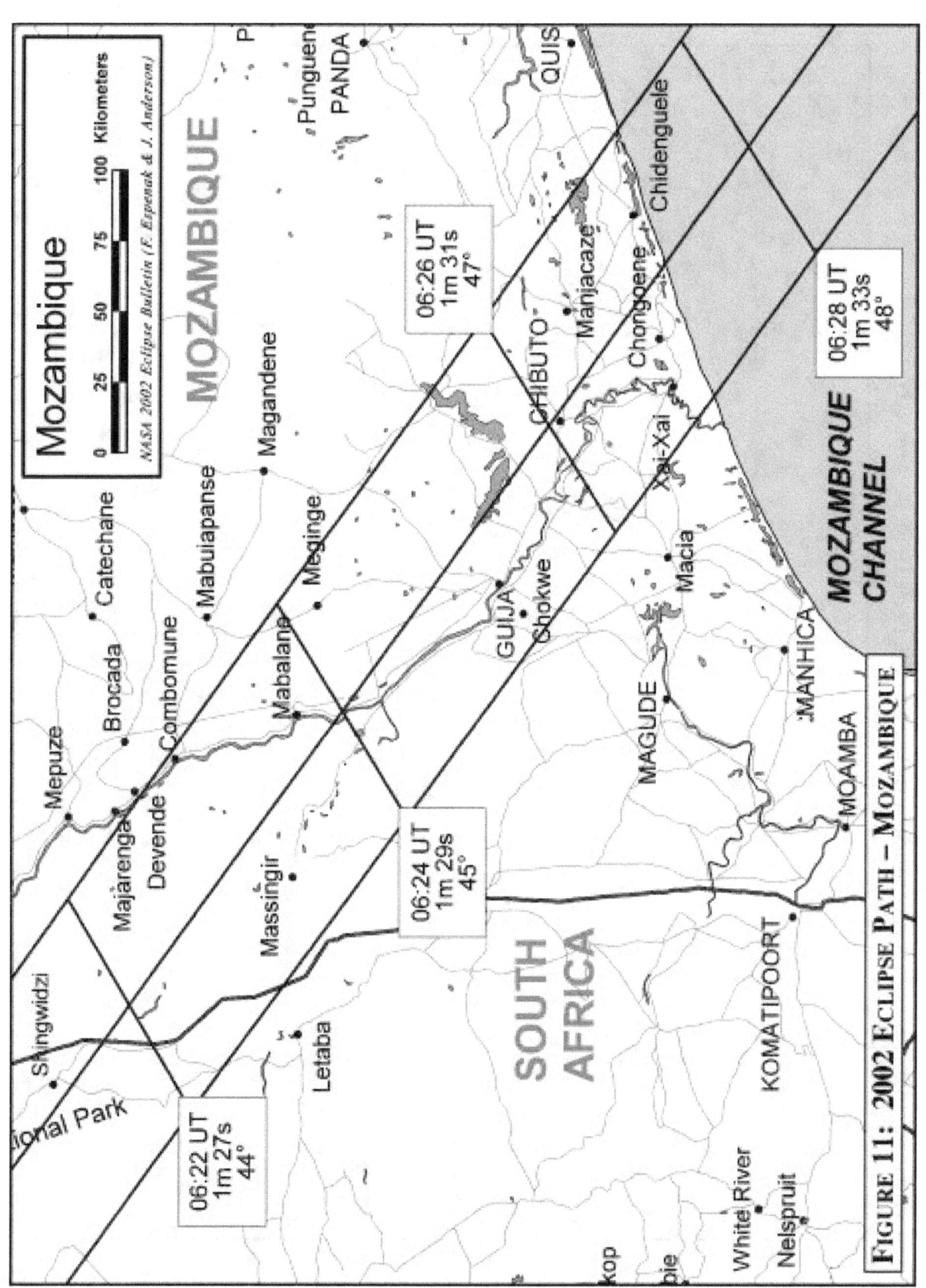

Figure 11: 2002 Eclipse Path – Mozambique

Total Solar Eclipse of 2002 December 04

Figure 12: The Eclipse Path Through Australia

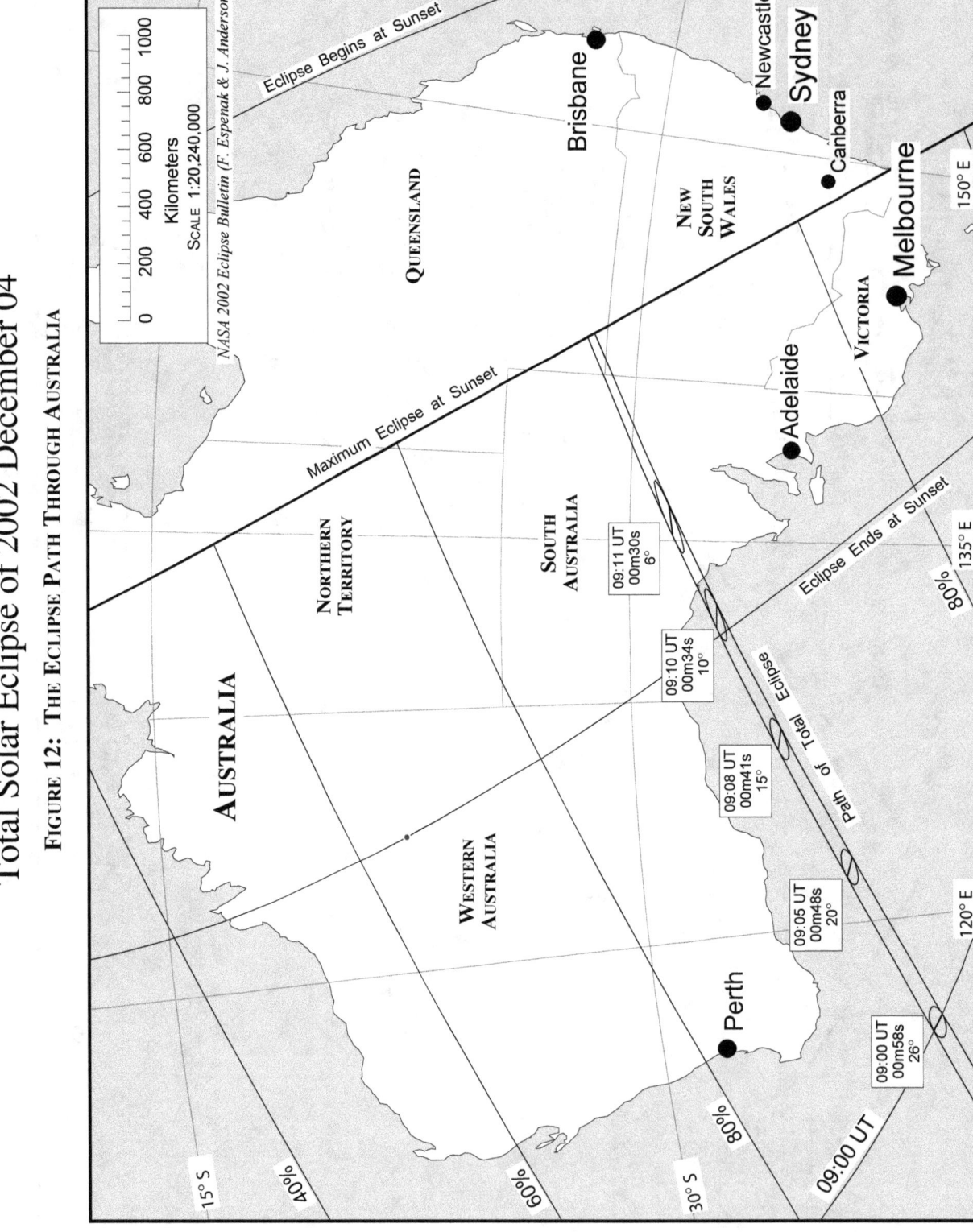

Total Solar Eclipse of 2002 December 04

Figure 13: The Eclipse Path Through Southern Australia

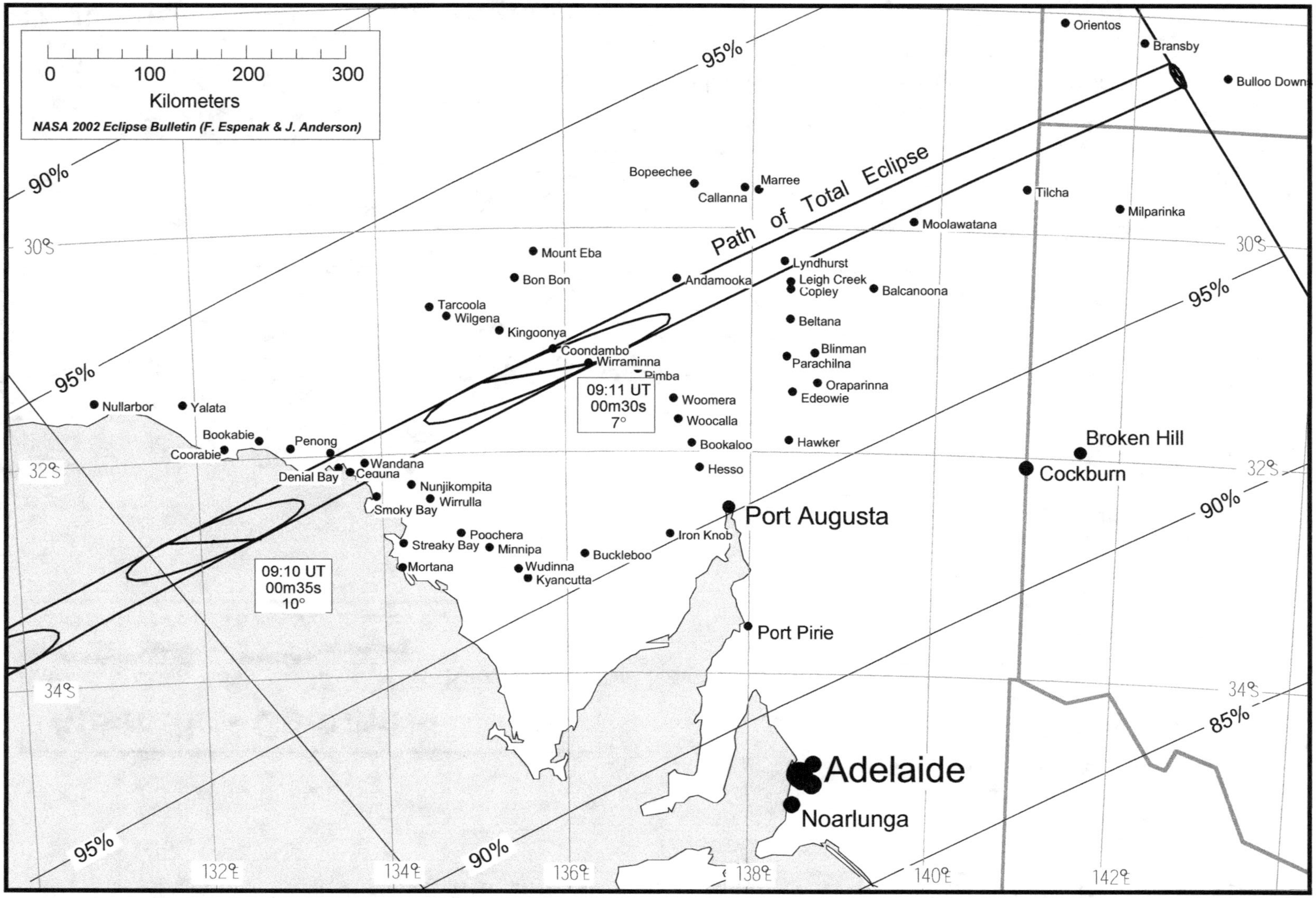

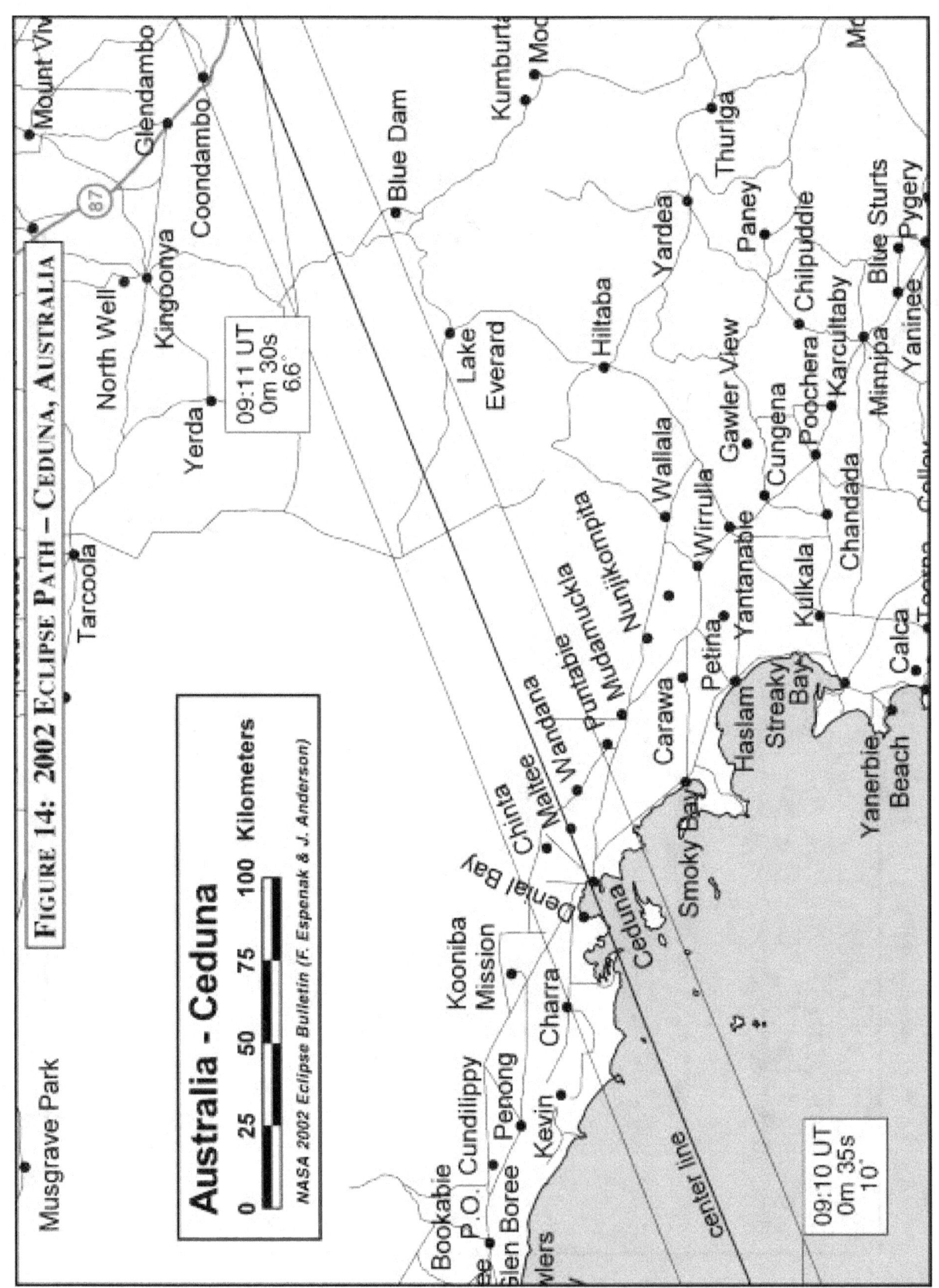

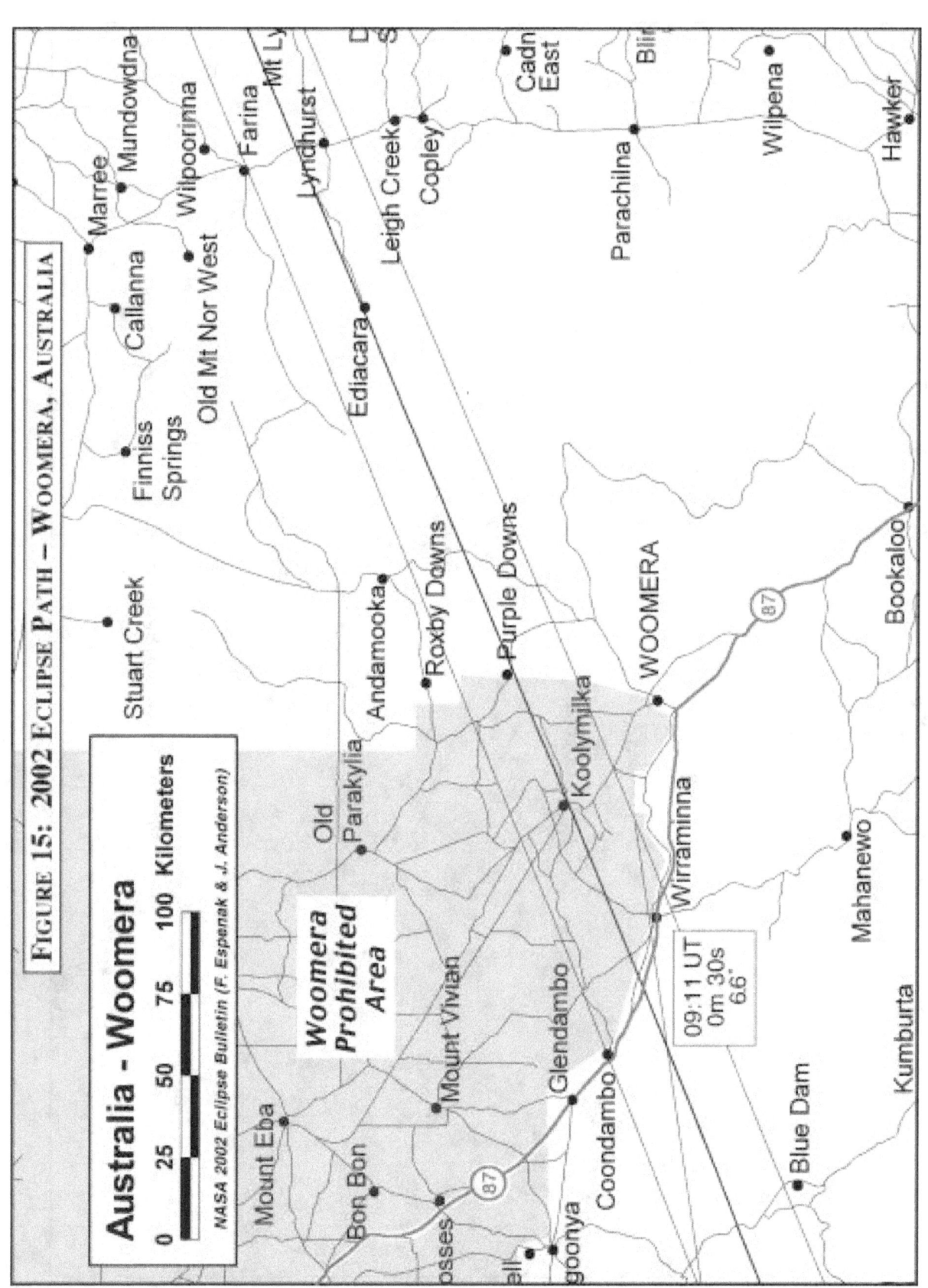

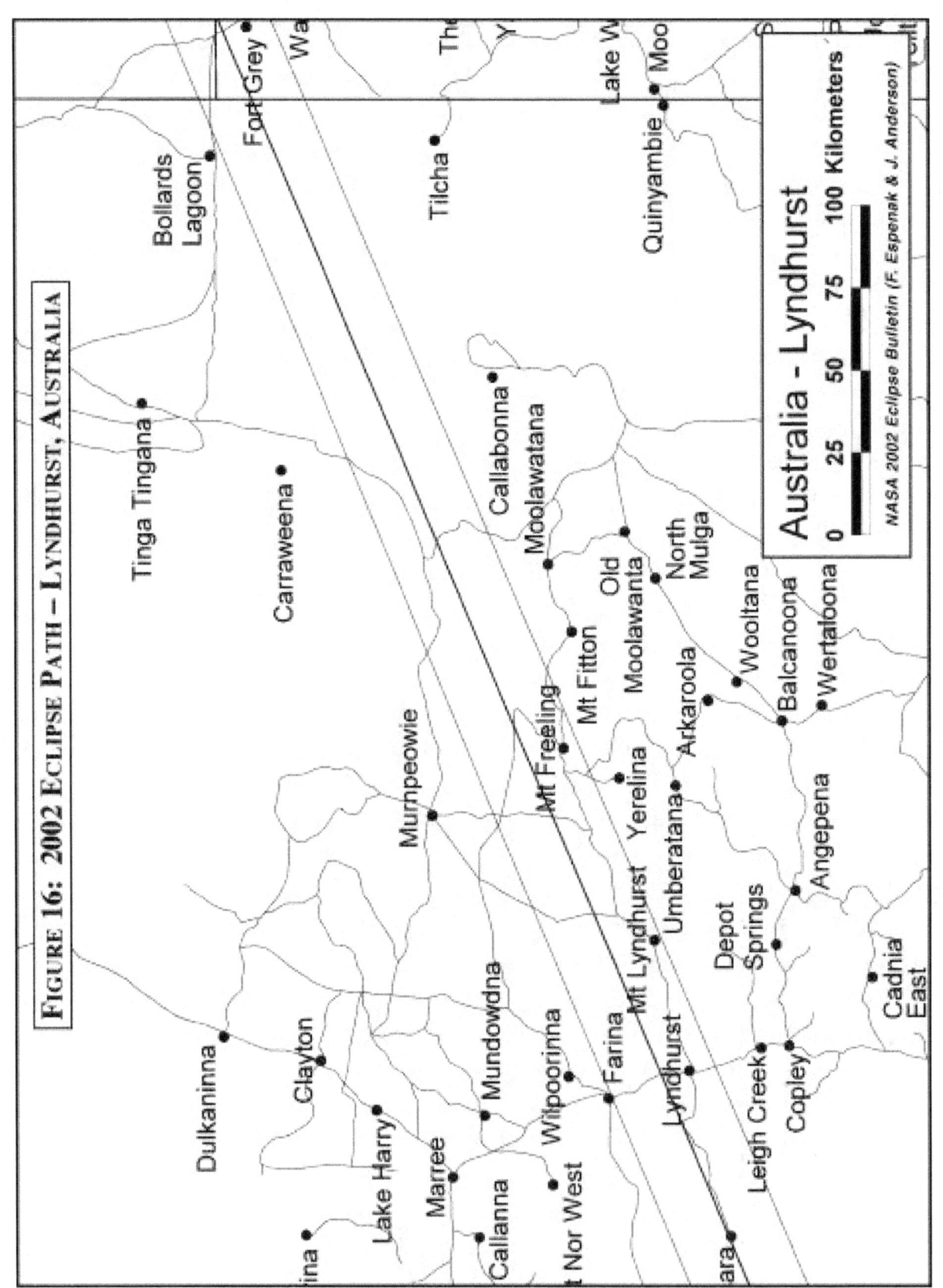

Figure 16: 2002 Eclipse Path – Lyndhurst, Australia

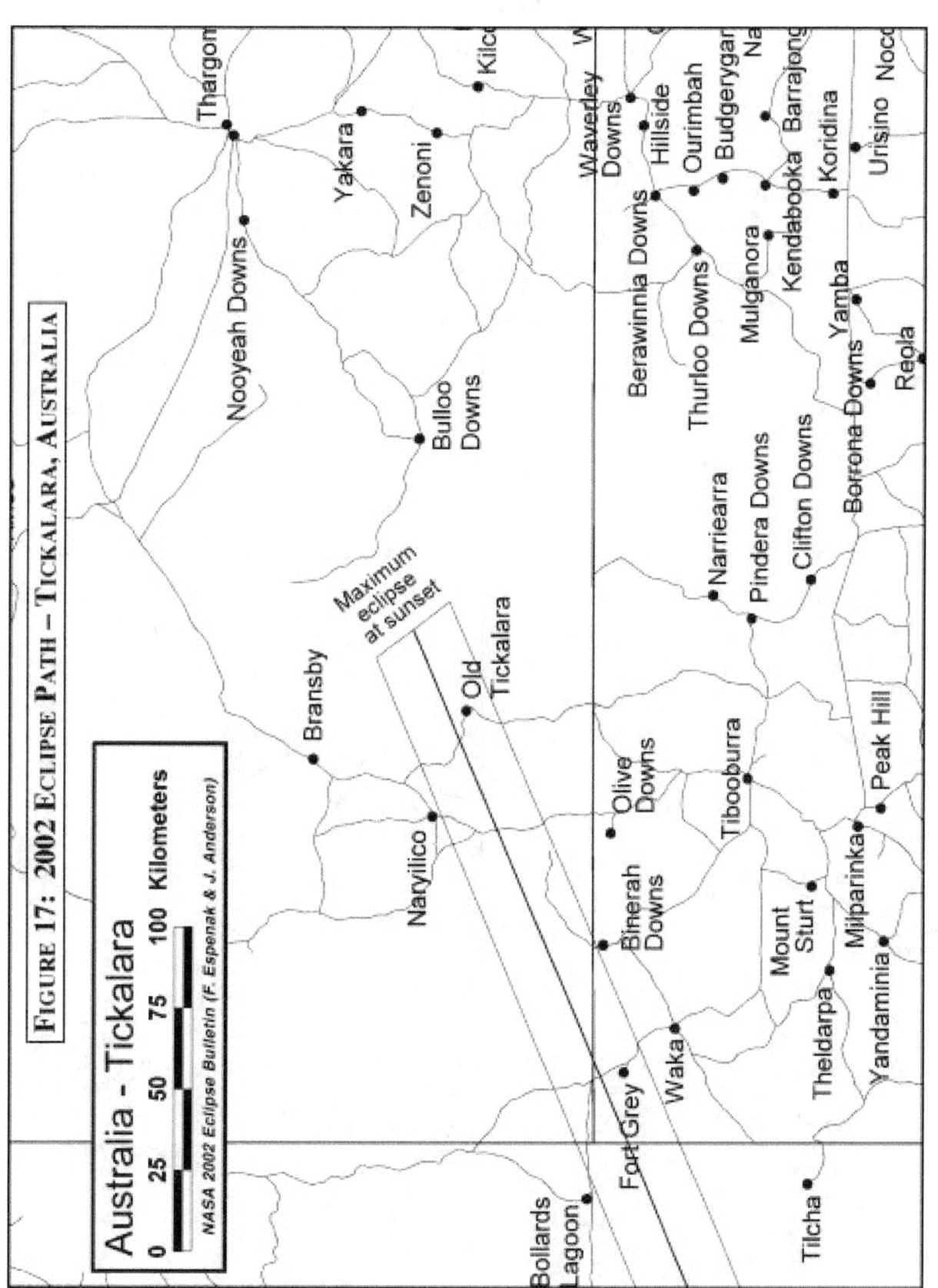

Figure 17: 2002 Eclipse Path – Tickalara, Australia

Total Solar Eclipse of 2002 Dec 04

Figure 18: The Lunar Limb Profile at 06:15 UT

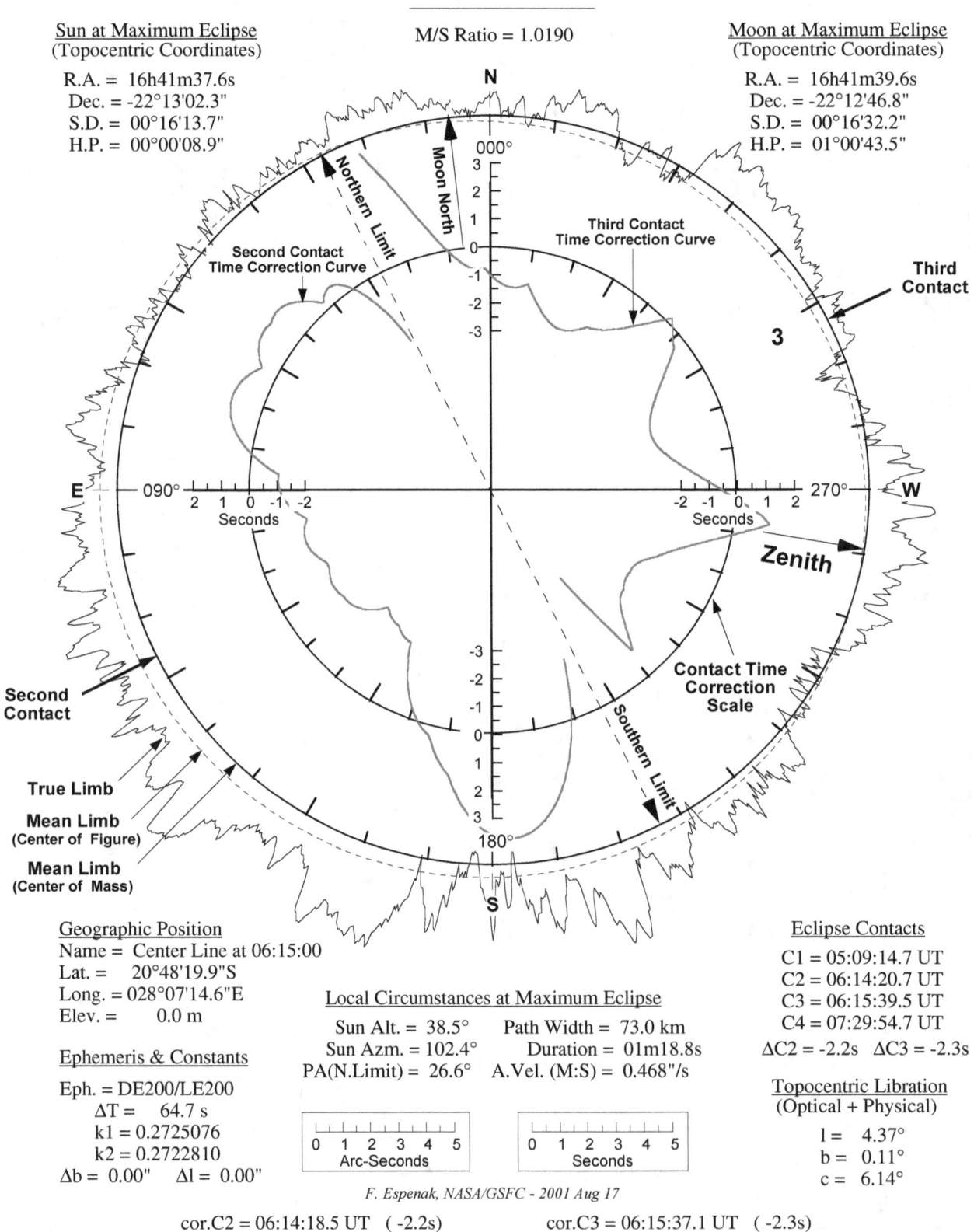

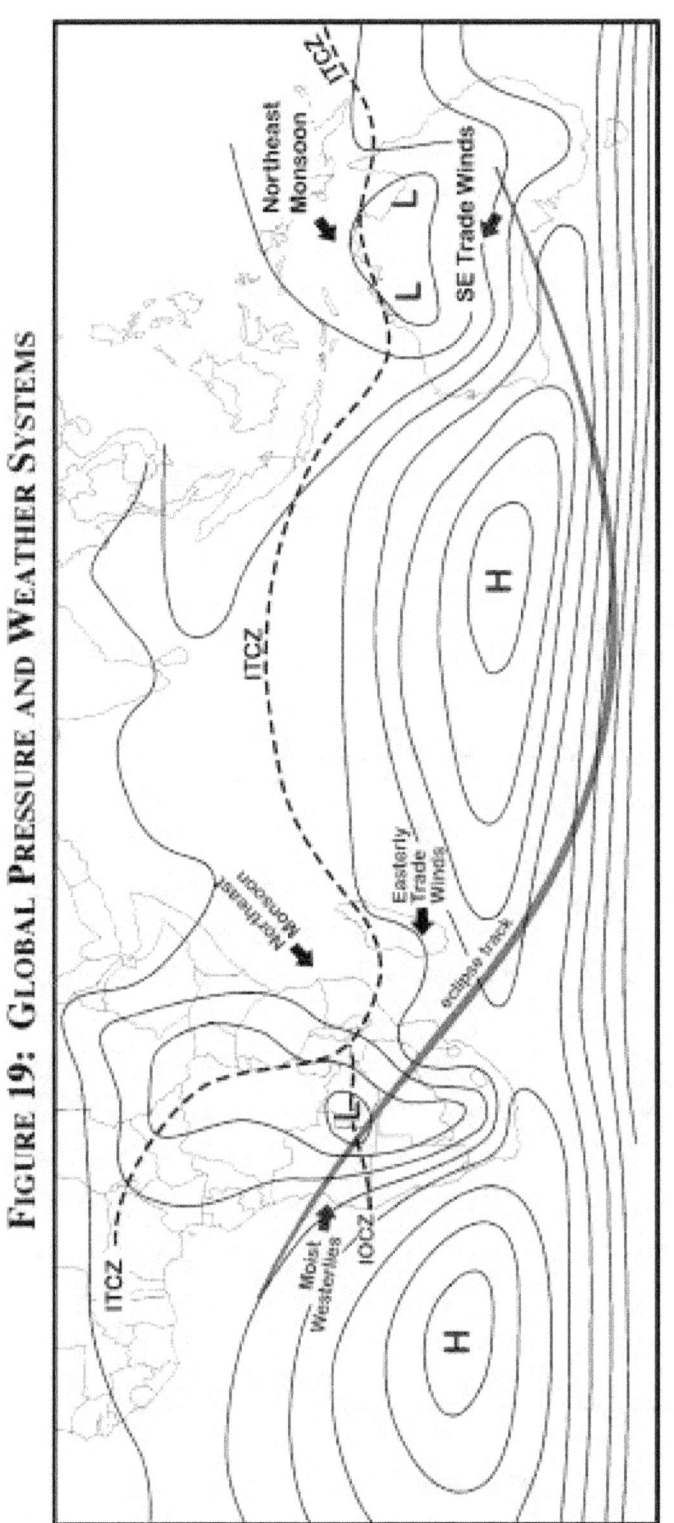

FIGURE 19: GLOBAL PRESSURE AND WEATHER SYSTEMS

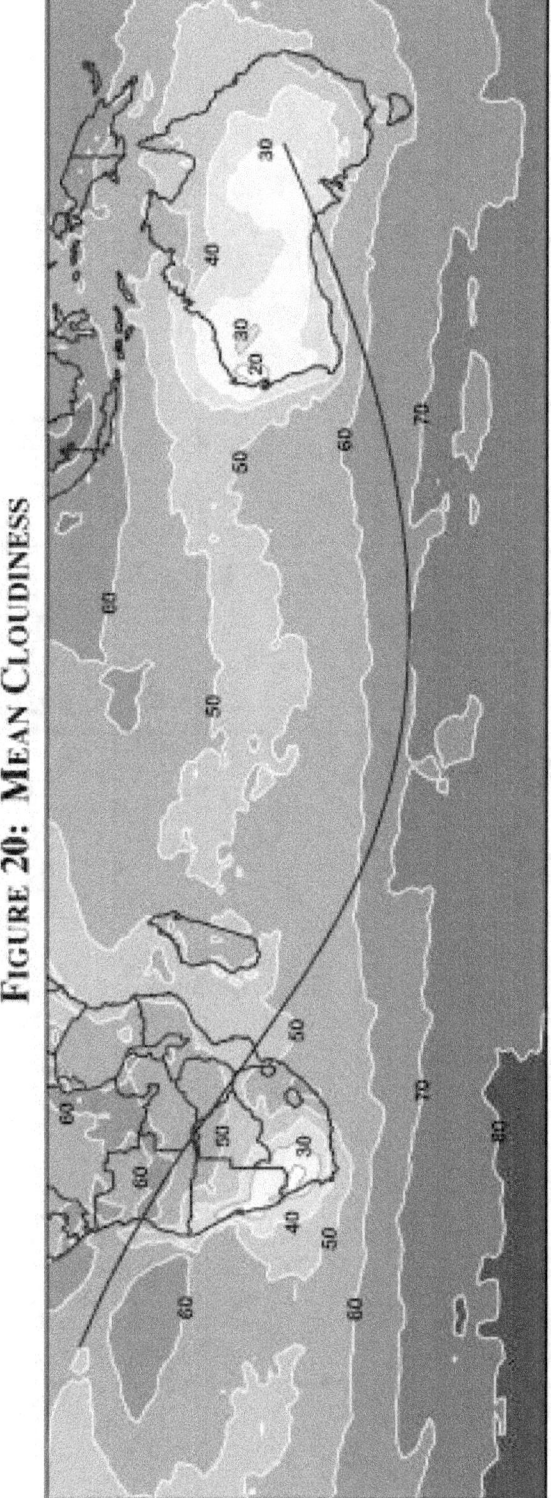

FIGURE 20: MEAN CLOUDINESS

FIGURE 21: MEAN CLOUDINESS ALONG THE ECLIPSE PATH

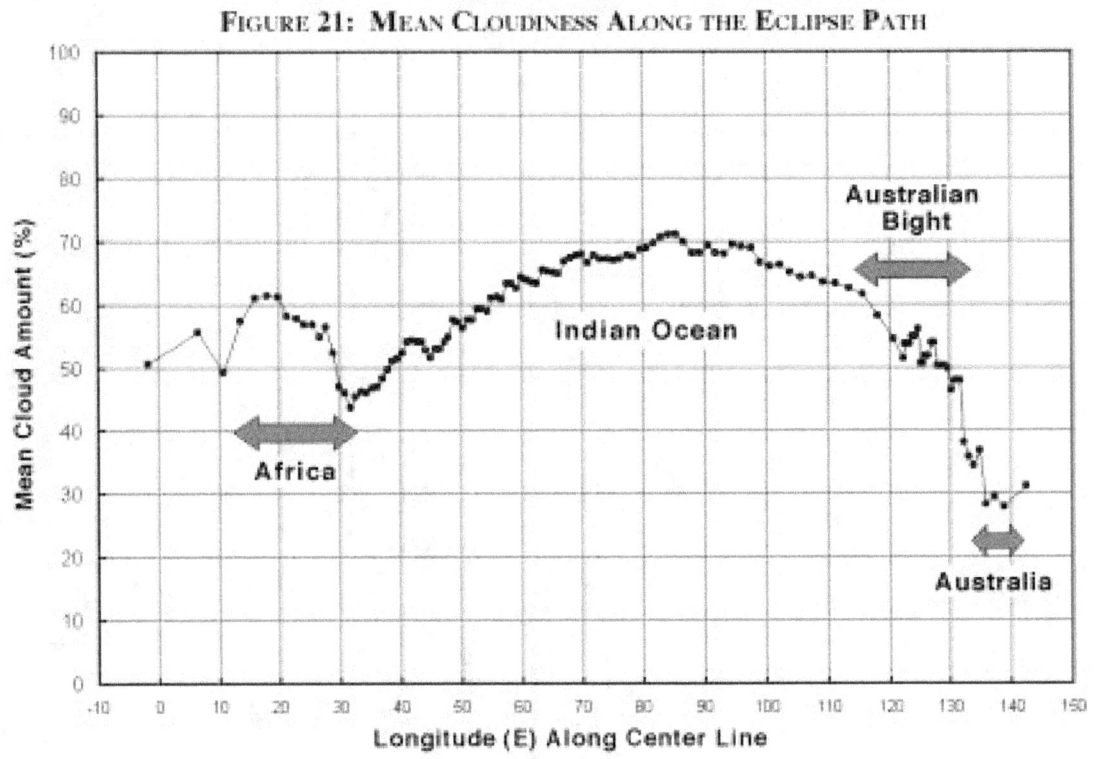

FIGURE 22: CLEAR SKIES IN AUSTRALIA

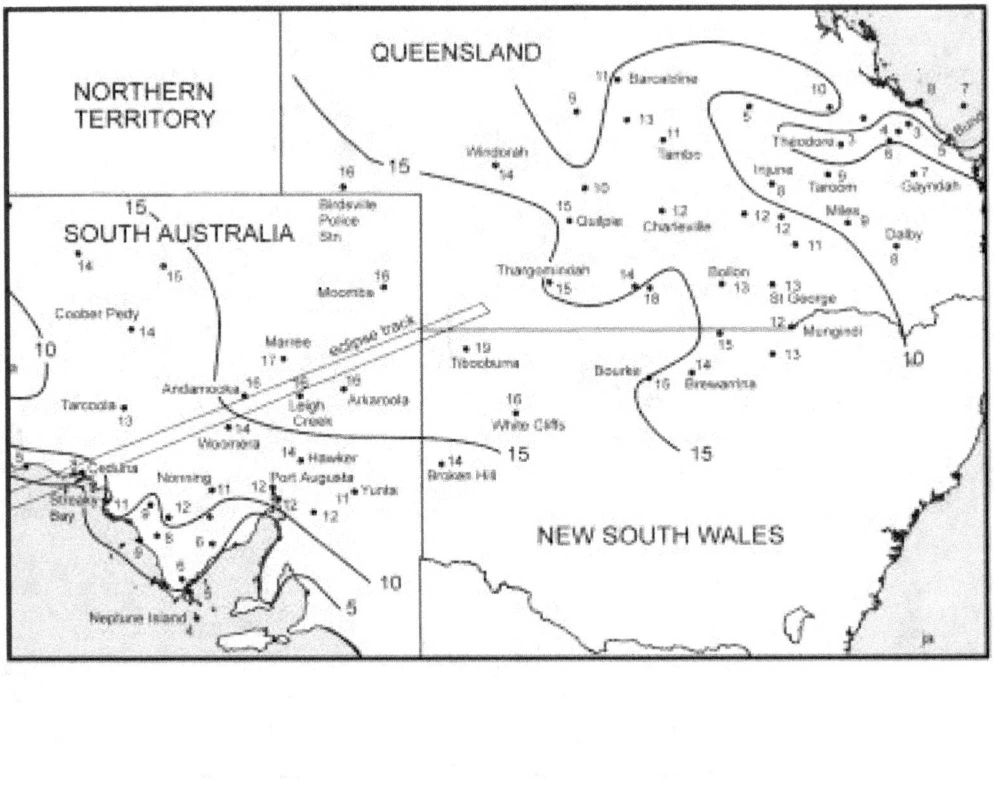

FIGURE 23: SPECTRAL RESPONSE OF SOME COMMONLY AVAILABLE SOLAR FILTERS

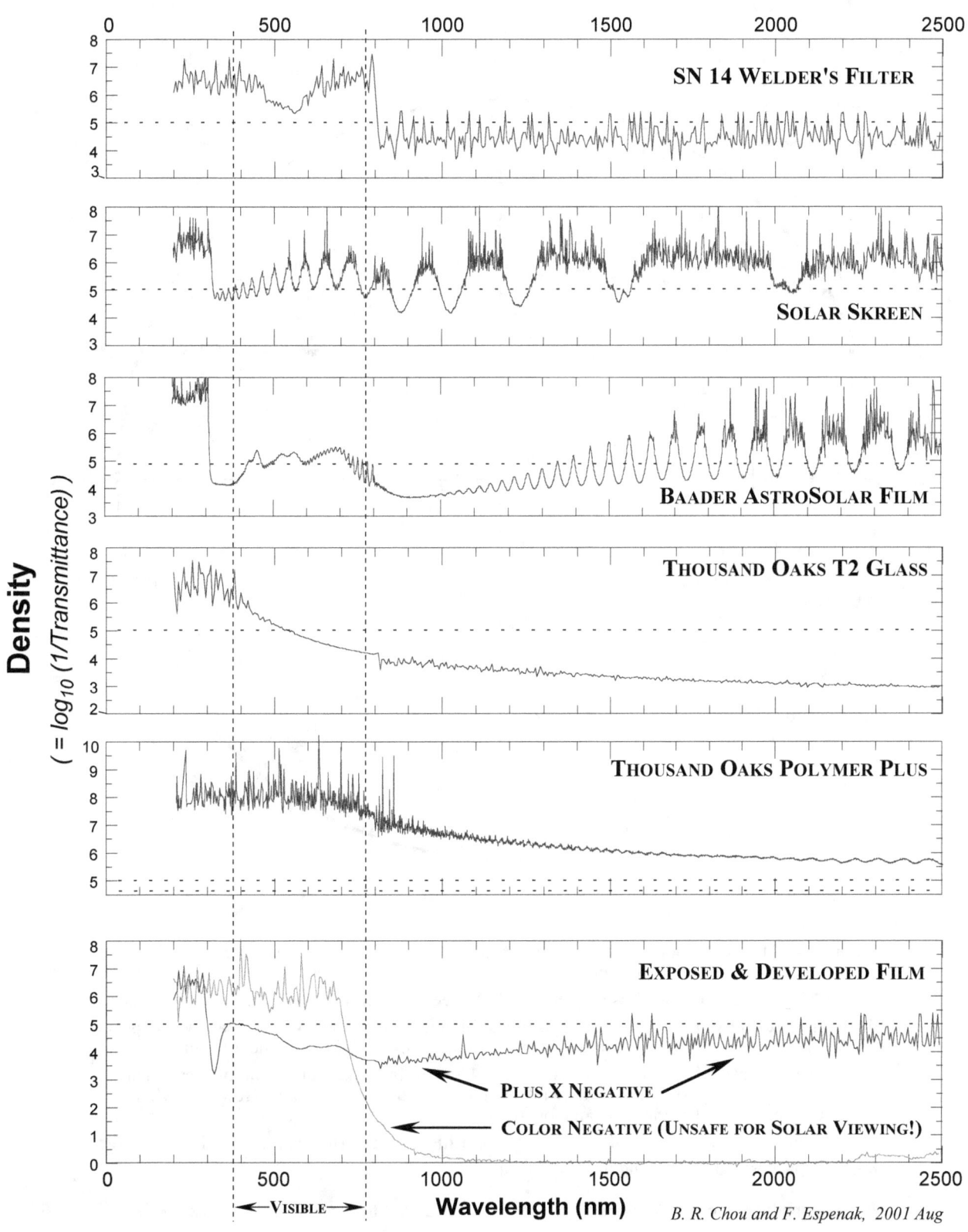

Total Solar Eclipse of 2002 Dec 04

Figure 24: The Sky During Totality As Seen From Center Line At 06:15 UT

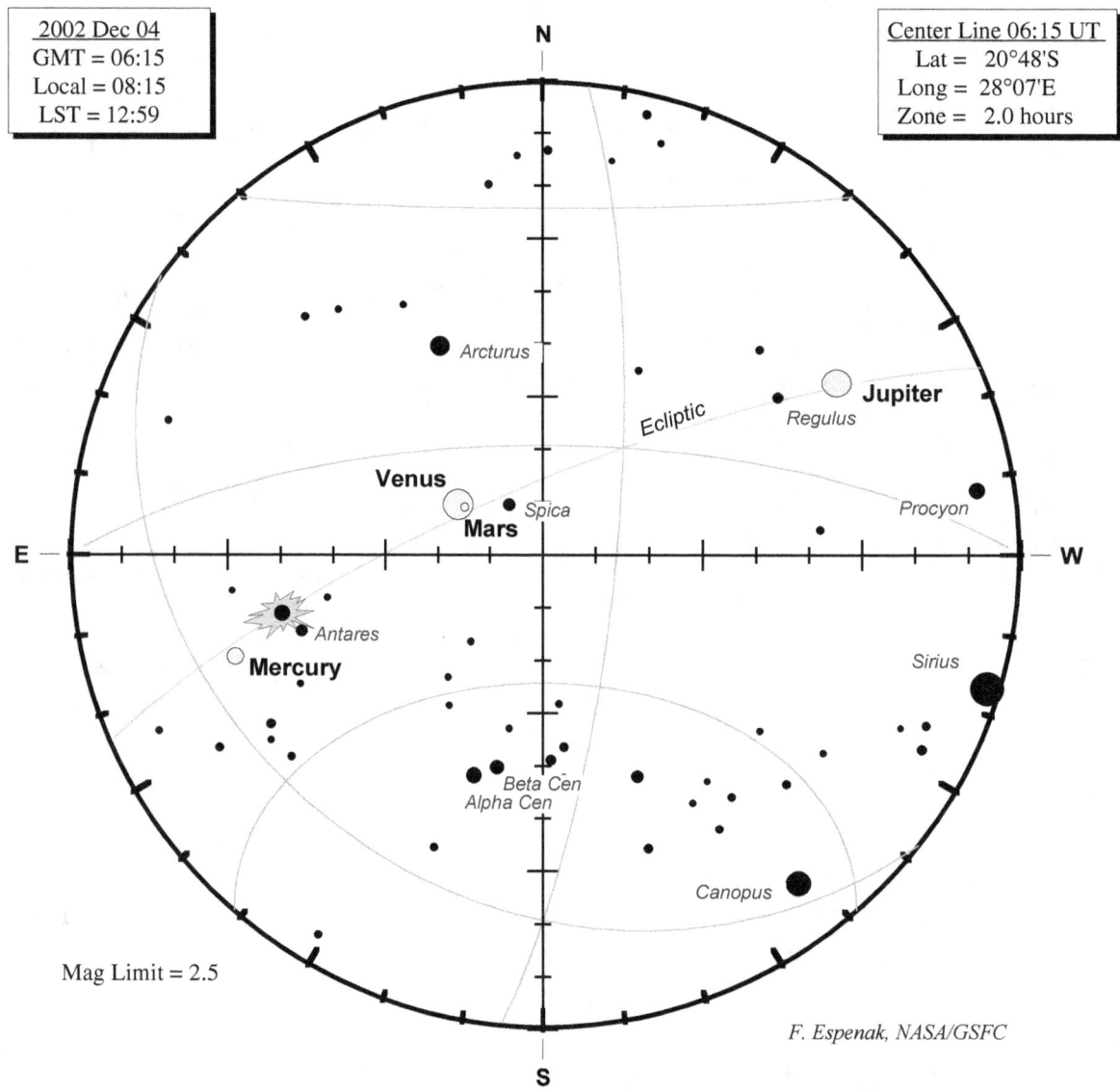

Figure 24: The sky during totality as seen from the center line Zimbabwe at 06:15 UT. The most conspicuous planet visible during totality will be Venus ($m_v=-4.5$) located 39° west of the Sun in Virgo. Mars ($m_v=+1.7$) is one and a half degrees west of Venus, but offers a much more difficult target since it will be nearly 500 times fainter. From Zimbabwe, the pair will appear 70° high in the northeast. Compared to Venus, Jupiter ($m_v=-2.2$) is the next brightest planet but it will be located in the northwestern sky 114° away from the Sun. None of these planets will be visible from Australia since they all set hours before totality begins. However, Mercury ($m_v=-0.6$) should be discernible from most places along the eclipse track. The innermost planet lies 11° east of the Sun. Since it is nearly at opposition, Saturn ($m_v=-0.1$) will be below the horizon for most locations along the umbral path.

For sky maps from other locations along the eclipse path, see the special 2002 eclipse web site:

http://sunearth.gsfc.nasa.gov/eclipse/TSE2002T/SE2002.html

Table 1

Elements of the Total Solar Eclipse of 2002 December 04

<u>Geocentric Conjunction</u>   07:39:48.73 TDT   J.D. = 2452612.819314
<u>of Sun & Moon in R.A.</u>:   (=07:38:44.03 UT)

<u>Instant of</u>   07:32:15.65 TDT   J.D. = 2452612.814070
<u>Greatest Eclipse</u>:   (=07:31:10.95 UT)

<u>Geocentric Coordinates of Sun & Moon at Greatest Eclipse (DE200/LE200)</u>:

<u>Sun</u>:   R.A. = 16h41m50.939s   <u>Moon</u>:   R.A. = 16h41m32.881s
             Dec. =-22°13'29.20"   Dec. =-22°31'05.17"
    Semi-Diameter =   16'13.60"   Semi-Diameter =   16'21.60"
      Eq.Hor.Par. =   08.92"   Eq.Hor.Par. = 1°00'02.27"
          Δ R.A. =   10.867s/h Δ R.A. =   154.401s/h
          Δ Dec. =   -20.23"/h Δ Dec. =   -487.39"/h

<u>Lunar Radius</u> k_1 = 0.2725076 (Penumbra)   <u>Shift in</u> Δb = 0.00"
<u>Constants</u>: k_2 = 0.2722810 (Umbra)   <u>Lunar Position</u>: Δl = 0.00"

<u>Geocentric Libration</u>:   l =   3.8°   Brown Lun. No. = 989
(Optical + Physical)   b =   0.5°   Saros Series   = 142 (22/72)
                         c =   5.7°   Ephemeris   = (DE200/LE200)

<u>Eclipse Magnitude</u> = 1.02437   <u>Gamma</u> =-0.30204 ΔT =   64.7 s

<u>Polynomial Besselian Elements for</u>:   2002 Dec 04   08:00:00.0 TDT   (=t_0)

n	x	y	d	l_1	l_2	µ
0	0.1861473	-0.3544635	-22.2264977	0.5441862	-0.0019615	302.485046
1	0.5532511	-0.1309133	-0.0052963	0.0000834	0.0000830	14.997272
2	0.0000155	0.0001807	0.0000058	-0.0000125	-0.0000124	-0.000002
3	-0.0000087	0.0000022	0.0000000	0.0000000	0.0000000	0.000000

Tan f_1 = 0.0047437   Tan f_2 = 0.0047201

At time 't_1' (decimal hours), each Besselian element is evaluated by:

$$a = a_0 + a_1 \ast t + a_2 \ast t^2 + a_3 \ast t^3 \quad \text{(or } a = \Sigma\ [a_n \ast t^n];\ n = 0 \text{ to } 3)$$

where: a = x, y, d, l_1, l_2, or µ
         t = t_1 - t_0 (decimal hours) and t_0 = 8.000 TDT

The Besselian elements were derived from a least-squares fit to elements calculated at five uniformly spaced times over a six hour period centered at t_0. Thus the elements are valid over the period 5.00 ≤ t_1 ≤ 11.00 TDT.

Note that all times are expressed in Terrestrial Dynamical Time (TDT).

Saros Series 142: Member 22 of 72 eclipses in series.

TABLE 2

SHADOW CONTACTS AND CIRCUMSTANCES
TOTAL SOLAR ECLIPSE OF 2002 DECEMBER 04

$$\Delta T = 64.7 \text{ s}$$
$$= 0°16'13.2''$$

		Terrestrial Dynamical Time h m s	Latitude	Ephemeris Longitude†	True Longitude*
External/Internal Contacts of Penumbra:	P1	04:52:27.3	01°57.0'N	015°11.4'E	015°27.6'E
	P2	06:56:18.3	26°05.1'S	028°06.1'W	027°49.8'W
	P3	08:08:01.1	49°28.1'S	174°03.7'E	174°19.9'E
	P4	10:12:05.3	22°44.6'S	124°22.1'E	124°38.3'E
Extreme North/South Limits of Penumbral Path:	N1	05:39:35.8	24°51.3'N	013°30.9'E	013°47.2'E
	S1	06:40:34.2	40°30.0'S	033°03.0'W	032°46.8'W
	N2	09:25:02.8	00°12.2'N	126°05.3'E	126°21.5'E
	S2	08:23:41.4	61°14.5'S	170°16.7'W	170°00.5'W
External/Internal Contacts of Umbra:	U1	05:51:23.9	03°53.5'S	001°55.9'W	001°39.7'W
	U2	05:51:53.1	03°58.0'S	002°05.1'W	001°48.9'W
	U3	09:12:35.8	28°32.3'S	142°12.8'E	142°29.0'E
	U4	09:13:00.8	28°28.6'S	142°04.5'E	142°20.7'E
Extreme North/South Limits of Umbral Path:	N1	05:51:32.0	03°48.5'S	001°55.9'W	001°39.7'W
	S1	05:51:45.0	04°03.0'S	002°05.1'W	001°48.9'W
	N2	09:12:53.9	28°24.4'S	142°04.0'E	142°20.2'E
	S2	09:12:42.6	28°36.5'S	142°13.4'E	142°29.6'E
Extreme Limits of Center Line:	C1	05:51:38.5	03°55.8'S	002°00.5'W	001°44.3'W
	C2	09:12:48.3	28°30.5'S	142°08.7'E	142°24.9'E
Instant of Greatest Eclipse:	G0	07:32:15.7	39°27.4'S	059°16.9'E	059°33.1'E
Circumstances at Greatest Eclipse:	Sun's Altitude = 72.2° Path Width = 87.0 km				
	Sun's Azimuth = 15.8° Central Duration = 02m03.8s				

† Ephemeris Longitude is the terrestrial dynamical longitude assuming a uniformly rotating Earth.
* True Longitude is calculated by correcting the Ephemeris Longitude for the non-uniform rotation of Earth.
   (T.L. = E.L. + 1.002738*ΔT/240, where ΔT(in seconds) = TDT - UT)

Note: Longitude is measured positive to the East.

Since ΔT is not known in advance, the value used in the predictions is an extrapolation based on pre-2002 measurements. Nevertheless, the actual value is expected to fall within ±0.5 seconds of the estimated ΔT used here.

TABLE 3

PATH OF THE UMBRAL SHADOW
TOTAL SOLAR ECLIPSE OF 2002 DECEMBER 04

Universal Time	Northern Limit Latitude	Northern Limit Longitude	Southern Limit Latitude	Southern Limit Longitude	Center Line Latitude	Center Line Longitude	Sun Alt °	Path Width km	Central Durat.
Limits	03°48.5'S	001°39.7'W	04°03.0'S	001°48.9'W	03°55.8'S	001°44.3'W	0	31	00m26.1s
05:55	10°33.8'S	012°31.9'E	10°38.8'S	011°43.8'E	10°36.4'S	012°07.9'E	16	51	00m46.5s
06:00	13°51.3'S	018°11.4'E	14°02.6'S	017°28.1'E	13°57.1'S	017°49.8'E	24	59	00m57.0s
06:05	16°26.3'S	022°16.2'E	16°41.4'S	021°34.3'E	16°33.9'S	021°55.3'E	29	65	01m05.3s
06:10	18°39.3'S	025°35.6'E	18°57.3'S	024°54.2'E	18°48.4'S	025°15.0'E	34	69	01m12.4s
06:15	20°38.0'S	028°27.7'E	20°58.5'S	027°46.7'E	20°48.3'S	028°07.2'E	39	73	01m18.8s
06:20	22°26.3'S	031°01.8'E	22°49.1'S	030°21.0'E	22°37.8'S	030°41.5'E	42	76	01m24.6s
06:25	24°06.5'S	033°23.4'E	24°31.3'S	032°42.8'E	24°19.0'S	033°03.2'E	46	78	01m29.8s
06:30	25°40.1'S	035°35.9'E	26°06.8'S	034°55.6'E	25°53.6'S	035°15.8'E	49	80	01m34.7s
06:35	27°08.1'S	037°41.8'E	27°36.6'S	037°02.0'E	27°22.4'S	037°22.0'E	52	82	01m39.1s
06:40	28°31.1'S	039°43.1'E	29°01.3'S	039°03.7'E	28°46.3'S	039°23.5'E	55	83	01m43.2s
06:45	29°49.7'S	041°41.0'E	30°21.5'S	041°02.3'E	30°05.7'S	041°21.8'E	58	84	01m46.9s
06:50	31°04.3'S	043°36.8'E	31°37.6'S	042°58.9'E	31°21.0'S	043°17.9'E	60	85	01m50.2s
06:55	32°15.1'S	045°31.4'E	32°49.8'S	044°54.4'E	32°32.5'S	045°13.0'E	63	86	01m53.2s
07:00	33°22.3'S	047°25.5'E	33°58.4'S	046°49.7'E	33°40.4'S	047°07.7'E	65	86	01m55.9s
07:05	34°26.1'S	049°19.9'E	35°03.5'S	048°45.4'E	34°44.8'S	049°02.7'E	67	87	01m58.1s
07:10	35°26.5'S	051°15.2'E	36°05.2'S	050°42.1'E	35°45.9'S	050°58.8'E	69	87	02m00.0s
07:15	36°23.6'S	053°11.9'E	37°03.5'S	052°40.6'E	36°43.6'S	052°56.3'E	70	87	02m01.5s
07:20	37°17.5'S	055°10.6'E	37°58.4'S	054°41.1'E	37°38.0'S	054°55.9'E	71	87	02m02.7s
07:25	38°08.0'S	057°11.6'E	38°50.0'S	056°44.3'E	38°29.0'S	056°58.0'E	72	87	02m03.4s
07:30	38°55.2'S	059°15.5'E	39°38.0'S	058°50.5'E	39°16.7'S	059°03.1'E	72	87	02m03.8s
07:35	39°39.0'S	061°22.7'E	40°22.6'S	061°00.3'E	40°00.8'S	061°11.5'E	72	87	02m03.7s
07:40	40°19.3'S	063°33.5'E	41°03.5'S	063°14.0'E	40°41.4'S	063°23.8'E	72	86	02m03.3s
07:45	40°55.9'S	065°48.4'E	41°40.5'S	065°31.9'E	41°18.2'S	065°40.2'E	71	86	02m02.4s
07:50	41°28.6'S	068°07.9'E	42°13.6'S	067°54.6'E	41°51.1'S	068°01.2'E	69	85	02m01.1s
07:55	41°57.4'S	070°32.1'E	42°42.5'S	070°22.3'E	42°19.9'S	070°27.2'E	68	84	01m59.5s
08:00	42°21.9'S	073°01.7'E	43°06.9'S	072°55.6'E	42°44.4'S	072°58.6'E	66	83	01m57.4s
08:05	42°41.8'S	075°37.1'E	43°26.5'S	075°34.7'E	43°04.2'S	075°35.8'E	64	82	01m54.9s
08:10	42°56.9'S	078°18.6'E	43°41.1'S	078°20.1'E	43°19.0'S	078°19.3'E	61	81	01m51.9s
08:15	43°06.8'S	081°06.9'E	43°50.2'S	081°12.5'E	43°28.5'S	081°09.6'E	59	80	01m48.6s
08:20	43°11.0'S	084°02.6'E	43°53.4'S	084°12.2'E	43°32.2'S	084°07.3'E	56	79	01m44.9s
08:25	43°09.0'S	087°06.5'E	43°50.1'S	087°20.2'E	43°29.5'S	087°13.2'E	54	77	01m40.7s
08:30	43°00.1'S	090°19.6'E	43°39.5'S	090°37.3'E	43°19.8'S	090°28.3'E	51	75	01m36.1s
08:35	42°43.5'S	093°43.3'E	43°20.9'S	094°04.9'E	43°02.2'S	093°53.9'E	47	73	01m31.1s
08:40	42°18.0'S	097°19.4'E	42°53.1'S	097°44.8'E	42°35.6'S	097°31.9'E	44	70	01m25.6s
08:45	41°42.2'S	101°10.7'E	42°14.6'S	101°39.9'E	41°58.4'S	101°25.1'E	40	68	01m19.7s
08:50	40°53.8'S	105°21.7'E	41°22.9'S	105°54.3'E	41°08.4'S	105°37.8'E	36	64	01m13.2s
08:55	39°49.3'S	109°59.4'E	40°14.5'S	110°35.4'E	40°02.0'S	110°17.2'E	32	60	01m06.0s
09:00	38°22.3'S	115°17.7'E	38°42.6'S	115°57.2'E	38°32.6'S	115°37.3'E	26	56	00m57.8s
09:05	36°18.5'S	121°48.5'E	36°32.2'S	122°32.6'E	36°25.5'S	122°10.4'E	20	49	00m48.2s
09:10	32°43.3'S	131°39.0'E	32°42.7'S	132°38.5'E	32°43.3'S	132°08.4'E	10	38	00m34.6s
Limits	28°24.4'S	142°20.2'E	28°36.5'S	142°29.6'E	28°30.5'S	142°24.9'E	0	27	00m22.4s

TABLE 4

PHYSICAL EPHEMERIS OF THE UMBRAL SHADOW
TOTAL SOLAR ECLIPSE OF 2002 DECEMBER 04

Universal Time	Center Line Latitude	Center Line Longitude	Diameter Ratio	Eclipse Obscur.	Sun Alt °	Sun Azm °	Path Width km	Major Axis km	Minor Axis km	Umbra Veloc. km/s	Central Durat.
05:50.6	03°55.8'S	001°44.3'W	1.0081	1.0163	0.0	112.3	31.3	-	28.0	-	00m26.1s
05:55	10°36.4'S	012°07.9'E	1.0130	1.0262	16.3	110.2	50.8	159.9	44.7	0.545	00m46.5s
06:00	13°57.1'S	017°49.8'E	1.0151	1.0305	23.8	108.4	59.2	129.1	52.0	0.518	00m57.0s
06:05	16°33.9'S	021°55.3'E	1.0167	1.0337	29.5	106.6	65.0	116.4	57.2	0.498	01m05.3s
06:10	18°48.4'S	025°15.0'E	1.0179	1.0362	34.3	104.6	69.4	109.0	61.3	0.482	01m12.4s
06:15	20°48.3'S	028°07.2'E	1.0190	1.0383	38.5	102.4	73.0	104.1	64.8	0.468	01m18.8s
06:20	22°37.8'S	030°41.5'E	1.0198	1.0400	42.4	100.0	75.9	100.6	67.7	0.457	01m24.6s
06:25	24°19.0'S	033°03.2'E	1.0206	1.0416	45.9	97.5	78.3	97.9	70.3	0.446	01m29.8s
06:30	25°53.6'S	035°15.8'E	1.0212	1.0429	49.2	94.7	80.2	95.8	72.5	0.437	01m34.7s
06:35	27°22.4'S	037°22.0'E	1.0218	1.0441	52.2	91.5	81.9	94.2	74.4	0.429	01m39.1s
06:40	28°46.3'S	039°23.5'E	1.0223	1.0451	55.1	88.1	83.2	92.8	76.0	0.421	01m43.2s
06:45	30°05.7'S	041°21.8'E	1.0228	1.0460	57.8	84.2	84.3	91.6	77.5	0.415	01m46.9s
06:50	31°21.0'S	043°17.9'E	1.0231	1.0468	60.4	79.9	85.2	90.7	78.8	0.409	01m50.2s
06:55	32°32.5'S	045°13.0'E	1.0235	1.0475	62.7	75.0	85.9	89.8	79.8	0.403	01m53.2s
07:00	33°40.4'S	047°07.7'E	1.0237	1.0480	64.9	69.4	86.5	89.2	80.7	0.399	01m55.9s
07:05	34°44.8'S	049°02.7'E	1.0239	1.0485	66.9	63.1	86.9	88.6	81.5	0.395	01m58.1s
07:10	35°45.9'S	050°58.8'E	1.0241	1.0488	68.6	55.9	87.1	88.1	82.0	0.391	02m00.0s
07:15	36°43.6'S	052°56.3'E	1.0242	1.0491	70.1	47.7	87.3	87.8	82.5	0.389	02m01.5s
07:20	37°38.0'S	054°55.9'E	1.0243	1.0493	71.2	38.7	87.3	87.5	82.7	0.386	02m02.7s
07:25	38°29.0'S	056°58.0'E	1.0244	1.0493	71.9	28.8	87.2	87.2	82.9	0.385	02m03.4s
07:30	39°16.7'S	059°03.1'E	1.0244	1.0493	72.2	18.3	87.0	87.1	82.9	0.383	02m03.8s
07:35	40°00.8'S	061°11.5'E	1.0243	1.0493	72.1	7.8	86.7	87.0	82.7	0.383	02m03.7s
07:40	40°41.4'S	063°23.8'E	1.0242	1.0491	71.5	357.4	86.3	87.0	82.5	0.383	02m03.3s
07:45	41°18.2'S	065°40.2'E	1.0241	1.0488	70.6	347.7	85.8	87.0	82.0	0.384	02m02.4s
07:50	41°51.1'S	068°01.2'E	1.0239	1.0485	69.3	338.7	85.1	87.1	81.5	0.385	02m01.1s
07:55	42°19.9'S	070°27.2'E	1.0237	1.0480	67.6	330.5	84.4	87.3	80.8	0.387	01m59.5s
08:00	42°44.4'S	072°58.6'E	1.0235	1.0475	65.8	323.0	83.5	87.6	79.9	0.390	01m57.4s
08:05	43°04.2'S	075°35.8'E	1.0232	1.0469	63.7	316.3	82.5	88.0	78.9	0.393	01m54.9s
08:10	43°19.0'S	078°19.3'E	1.0228	1.0461	61.4	310.1	81.3	88.5	77.7	0.397	01m51.9s
08:15	43°28.5'S	081°09.6'E	1.0224	1.0453	58.9	304.3	80.0	89.1	76.3	0.402	01m48.6s
08:20	43°32.2'S	084°07.3'E	1.0219	1.0443	56.3	299.0	78.5	89.9	74.7	0.407	01m44.9s
08:25	43°29.5'S	087°13.2'E	1.0214	1.0433	53.5	293.9	76.9	90.8	73.0	0.414	01m40.7s
08:30	43°19.8'S	090°28.3'E	1.0208	1.0420	50.5	289.1	75.0	92.0	71.0	0.421	01m36.1s
08:35	43°02.2'S	093°53.9'E	1.0201	1.0406	47.4	284.5	72.9	93.5	68.7	0.430	01m31.1s
08:40	42°35.6'S	097°31.9'E	1.0193	1.0391	43.9	279.9	70.4	95.4	66.1	0.440	01m25.6s
08:45	41°58.4'S	101°25.1'E	1.0185	1.0373	40.3	275.5	67.6	97.9	63.2	0.451	01m19.7s
08:50	41°08.4'S	105°37.8'E	1.0175	1.0352	36.2	271.0	64.4	101.3	59.8	0.465	01m13.2s
08:55	40°02.0'S	110°17.2'E	1.0163	1.0328	31.7	266.5	60.5	106.3	55.8	0.480	01m06.0s
09:00	38°32.6'S	115°37.3'E	1.0148	1.0299	26.5	261.8	55.6	114.4	50.9	0.499	00m57.8s
09:05	36°25.5'S	122°10.4'E	1.0130	1.0261	20.0	256.6	49.1	130.7	44.6	0.524	00m48.2s
09:10	32°43.3'S	132°08.4'E	1.0100	1.0201	10.1	250.0	38.4	197.7	34.5	0.564	00m34.6s
09:11.7	28°30.5'S	142°24.9'E	1.0069	1.0139	0.0	244.5	26.9	-	24.0	-	00m22.4s

TABLE 5

LOCAL CIRCUMSTANCES ON THE CENTER LINE
TOTAL SOLAR ECLIPSE OF 2002 DECEMBER 04

Center Line Maximum Eclipse			First Contact				Second Contact			Third Contact			Fourth Contact			
U.T.	Durat.	Alt °	U.T.	P °	V °	Alt °	U.T.	P °	V °	U.T.	P °	V °	U.T.	P °	V °	Alt °
05:55	00m46.5s	16	04:58:02	293	34	3	05:54:37	115	210	05:55:23	295	30	06:58:39	116	205	31
06:00	00m57.0s	24	05:00:15	294	35	10	05:59:32	115	211	06:00:29	296	31	07:07:20	117	207	39
06:05	01m05.3s	30	05:03:01	295	37	15	06:04:27	116	213	06:05:33	296	33	07:15:14	117	208	46
06:10	01m12.4s	34	05:06:03	295	38	20	06:09:24	116	215	06:10:36	296	34	07:22:43	117	210	51
06:15	01m18.8s	39	05:09:15	296	39	24	06:14:21	117	216	06:15:39	297	36	07:29:55	117	212	56
06:20	01m24.6s	42	05:12:35	296	41	27	06:19:18	117	218	06:20:42	297	37	07:36:52	117	214	60
06:25	01m29.8s	46	05:16:00	296	42	30	06:24:15	117	219	06:25:45	297	39	07:43:38	117	217	64
06:30	01m34.7s	49	05:19:31	296	43	33	06:29:13	117	221	06:30:47	297	41	07:50:13	117	221	67
06:35	01m39.1s	52	05:23:07	296	44	36	06:34:11	117	223	06:35:50	297	43	07:56:37	116	225	70
06:40	01m43.2s	55	05:26:47	296	46	39	06:39:09	116	225	06:40:52	296	45	08:02:52	116	232	73
06:45	01m46.9s	58	05:30:30	296	47	42	06:44:07	116	227	06:45:54	296	47	08:08:56	116	240	75
06:50	01m50.2s	60	05:34:18	296	48	44	06:49:05	116	230	06:50:55	296	50	08:14:51	115	251	77
06:55	01m53.2s	63	05:38:09	296	50	47	06:54:03	116	233	06:55:57	295	54	08:20:37	114	264	78
07:00	01m55.9s	65	05:42:04	295	51	49	06:59:02	115	237	07:00:58	295	58	08:26:13	114	278	78
07:05	01m58.1s	67	05:46:03	295	53	51	07:04:01	115	242	07:05:59	295	62	08:31:40	113	291	77
07:10	02m00.0s	69	05:50:06	295	55	54	07:09:00	114	247	07:11:00	294	67	08:36:58	112	302	76
07:15	02m01.5s	70	05:54:14	294	57	56	07:13:59	114	253	07:16:01	294	74	08:42:06	112	310	75
07:20	02m02.7s	71	05:58:26	294	60	58	07:18:59	113	260	07:21:01	293	81	08:47:07	111	316	73
07:25	02m03.4s	72	06:02:43	294	63	60	07:23:58	112	267	07:26:02	292	89	08:51:58	110	321	71
07:30	02m03.8s	72	06:07:06	293	66	62	07:28:58	112	276	07:31:02	292	97	08:56:42	109	324	68
07:35	02m03.7s	72	06:11:34	292	69	63	07:33:58	111	284	07:36:02	291	105	09:01:17	109	327	66
07:40	02m03.3s	72	06:16:08	292	74	65	07:38:58	110	292	07:41:02	290	113	09:05:45	108	329	64
07:45	02m02.4s	71	06:20:48	291	78	66	07:43:59	109	299	07:46:01	289	120	09:10:05	107	330	61
07:50	02m01.1s	69	06:25:36	291	84	68	07:48:59	109	305	07:51:01	289	126	09:14:18	106	331	59
07:55	01m59.5s	68	06:30:31	290	90	68	07:54:00	108	311	07:56:00	288	132	09:18:25	105	331	56
08:00	01m57.4s	66	06:35:34	289	97	69	07:59:01	107	315	08:00:59	287	136	09:22:24	105	332	54
08:05	01m54.9s	64	06:40:45	288	105	69	08:04:03	106	319	08:05:57	286	140	09:26:18	104	332	51
08:10	01m51.9s	61	06:46:06	287	112	69	08:09:04	105	322	08:10:56	285	143	09:30:06	103	332	49
08:15	01m48.6s	59	06:51:36	286	119	68	08:14:06	104	325	08:15:54	284	145	09:33:47	102	332	46
08:20	01m44.9s	56	06:57:17	286	126	67	08:19:07	103	327	08:20:52	283	147	09:37:23	101	332	43
08:25	01m40.7s	54	07:03:08	285	132	65	08:24:10	103	328	08:25:50	282	149	09:40:54	101	332	40
08:30	01m36.1s	51	07:09:12	284	138	63	08:29:12	102	330	08:30:48	282	150	09:44:19	100	332	37
08:35	01m31.1s	47	07:15:28	283	142	60	08:34:14	101	331	08:35:45	281	151	09:47:38	99	332	34
08:40	01m25.6s	44	07:21:58	282	146	57	08:39:17	100	332	08:40:43	280	152	09:50:50	99	331	31
08:45	01m19.7s	40	07:28:42	281	149	54	08:44:20	99	332	08:45:40	279	152	09:53:56	98	331	28
08:50	01m13.2s	36	07:35:43	280	152	50	08:49:23	98	333	08:50:37	278	153	09:56:54	97	331	24
08:55	01m06.0s	32	07:43:03	279	154	45	08:54:27	97	333	08:55:33	277	153	09:59:41	97	330	20
09:00	00m57.8s	27	07:50:48	277	155	40	08:59:31	97	333	09:00:29	277	153	10:02:14	96	330	15
09:05	00m48.2s	20	07:59:11	276	157	33	09:04:36	96	334	09:05:24	276	154	10:04:22	95	330	9
09:10	00m34.6s	10	08:09:06	275	158	22	09:09:43	95	334	09:10:17	275	154	10:05:23	95	329	0

Table 6

Topocentric Data and Path Corrections Due to Lunar Limb Profile
Total Solar Eclipse of 2002 December 04

Universal Time	Moon Topo H.P. "	Moon Topo S.D. "	Moon Rel. Ang.V "/s	Topo Lib. Long °	Sun Alt. °	Sun Az. °	Path Az. °	North Limit P.A. °	North Limit Int. '	North Limit Ext. '	South Limit Int. '	South Limit Ext. '	Central Durat. Corr. s
05:55	3622.0	986.4	0.545	4.54	16.3	110.2	119.5	24.6	0.2	1.4	0.1	-2.3	-0.3
06:00	3629.7	988.5	0.518	4.50	23.8	108.4	122.5	25.5	0.3	1.0	0.3	-2.3	-0.2
06:05	3635.3	990.0	0.498	4.46	29.5	106.6	124.5	26.0	0.3	0.6	0.3	-2.6	-0.2
06:10	3639.8	991.2	0.482	4.41	34.3	104.6	126.0	26.4	0.3	0.8	0.3	-2.7	-0.2
06:15	3643.5	992.2	0.468	4.37	38.5	102.4	127.0	26.6	0.3	0.9	0.3	-2.8	-0.1
06:20	3646.7	993.0	0.457	4.33	42.4	100.0	127.7	26.7	0.3	0.9	0.3	-2.8	-0.1
06:25	3649.4	993.8	0.446	4.29	45.9	97.5	128.1	26.8	0.3	0.9	0.3	-2.8	-0.6
06:30	3651.8	994.4	0.437	4.24	49.2	94.7	128.3	26.7	0.3	0.9	0.3	-2.8	-0.6
06:35	3653.9	995.0	0.429	4.20	52.2	91.5	128.2	26.6	0.3	0.9	0.2	-2.8	-0.6
06:40	3655.7	995.5	0.421	4.16	55.1	88.1	127.9	26.4	0.3	0.8	0.2	-2.7	-0.6
06:45	3657.3	995.9	0.415	4.12	57.8	84.2	127.4	26.2	0.3	0.7	0.2	-2.6	-0.6
06:50	3658.7	996.3	0.409	4.08	60.4	79.9	126.6	25.9	0.3	0.7	0.1	-2.4	-1.1
06:55	3659.8	996.6	0.403	4.03	62.7	75.0	125.8	25.5	0.3	0.9	0.1	-2.1	-1.1
07:00	3660.8	996.8	0.399	3.99	64.9	69.4	124.7	25.1	0.3	1.2	0.1	-2.2	-1.1
07:05	3661.6	997.0	0.395	3.95	66.9	63.1	123.5	24.6	0.3	1.3	0.0	-2.2	-1.1
07:10	3662.3	997.2	0.391	3.91	68.6	55.9	122.1	24.1	0.3	1.4	0.0	-2.1	-1.1
07:15	3662.7	997.3	0.389	3.87	70.1	47.7	120.5	23.5	0.3	1.5	-0.1	-1.9	-1.1
07:20	3663.1	997.4	0.386	3.82	71.2	38.7	118.9	22.9	0.3	1.4	-0.1	-2.0	-1.1
07:25	3663.2	997.5	0.385	3.78	71.9	28.8	117.0	22.3	0.3	1.2	-0.2	-2.6	-1.0
07:30	3663.2	997.5	0.383	3.74	72.2	18.3	115.1	21.6	0.4	1.0	-0.2	-3.1	-0.9
07:35	3663.1	997.4	0.383	3.70	72.1	7.8	113.0	20.9	0.3	1.1	-0.3	-3.4	-0.8
07:40	3662.8	997.3	0.383	3.65	71.5	357.4	110.8	20.1	0.4	1.2	-0.4	-3.6	-1.0
07:45	3662.3	997.2	0.384	3.61	70.6	347.7	108.5	19.4	0.4	1.0	-0.4	-3.6	-1.1
07:50	3661.7	997.1	0.385	3.57	69.3	338.7	106.1	18.6	0.4	0.9	-0.4	-3.4	-1.1
07:55	3660.9	996.8	0.387	3.52	67.6	330.5	103.6	17.7	0.4	0.8	-0.4	-3.5	-1.2
08:00	3660.0	996.6	0.390	3.48	65.8	323.0	101.1	16.9	0.5	1.1	-0.1	-3.5	-1.4
08:05	3658.8	996.3	0.393	3.44	63.7	316.3	98.4	16.0	0.5	1.1	0.0	-3.2	-1.4
08:10	3657.5	995.9	0.397	3.40	61.4	310.1	95.8	15.2	0.4	1.0	0.1	-2.8	-1.4
08:15	3656.1	995.5	0.402	3.36	58.9	304.3	93.0	14.3	0.3	1.2	0.2	-2.4	-1.3
08:20	3654.4	995.1	0.407	3.31	56.3	299.0	90.3	13.4	0.3	1.3	0.3	-2.6	-1.3
08:25	3652.4	994.6	0.414	3.27	53.5	293.9	87.5	12.5	0.2	1.2	0.3	-2.5	-1.5
08:30	3650.3	994.0	0.421	3.23	50.5	289.1	84.7	11.6	0.1	1.1	0.4	-2.2	-1.4
08:35	3647.8	993.3	0.430	3.19	47.4	284.5	81.9	10.8	0.1	0.9	0.4	-2.2	-1.3
08:40	3645.0	992.6	0.440	3.14	43.9	279.9	79.2	9.9	0.0	0.6	0.5	-2.5	-1.2
08:45	3641.8	991.7	0.451	3.10	40.3	275.5	76.5	9.0	0.0	0.8	0.5	-2.7	-1.0
08:50	3638.2	990.7	0.465	3.06	36.2	271.0	73.9	8.2	-0.0	0.8	0.6	-2.6	-0.9
08:55	3633.9	989.6	0.480	3.02	31.7	266.5	71.4	7.3	-0.0	0.8	0.6	-2.3	-0.8
09:00	3628.7	988.2	0.499	2.98	26.5	261.8	69.0	6.5	-0.1	0.7	0.6	-2.3	-0.8
09:05	3622.0	986.3	0.524	2.93	20.0	256.6	66.8	5.7	-0.1	0.8	0.7	-2.4	-0.8
09:10	3611.4	983.4	0.564	2.89	10.1	250.0	64.9	4.8	-0.3	0.6	0.7	-2.3	-0.9

Table 7
Mapping Coordinates for the Umbral Path – Africa
Total Solar Eclipse of 2002 December 04

Longitude	Latitude of:			Circumstances on Center Line				
	Northern Limit	Southern Limit	Center Line	Universal Time h m s	Sun Alt °	Sun Az. °	Path Width km	Central Durat.
013° 00.0'E	10° 49.40'S	11° 21.96'S	11° 05.60'S	05:55:37	17.4	110	52	00m48.0s
013° 30.0'E	11° 06.16'S	11° 39.27'S	11° 22.64'S	05:56:00	18.0	110	53	00m48.9s
014° 00.0'E	11° 23.09'S	11° 56.75'S	11° 39.84'S	05:56:24	18.7	110	54	00m49.8s
014° 30.0'E	11° 40.18'S	12° 14.40'S	11° 57.21'S	05:56:49	19.3	110	54	00m50.7s
015° 00.0'E	11° 57.44'S	12° 32.22'S	12° 14.74'S	05:57:15	20.0	109	55	00m51.6s
015° 30.0'E	12° 14.86'S	12° 50.20'S	12° 32.44'S	05:57:42	20.6	109	56	00m52.5s
016° 00.0'E	12° 32.44'S	13° 08.34'S	12° 50.31'S	05:58:09	21.3	109	56	00m53.5s
016° 30.0'E	12° 50.19'S	13° 26.65'S	13° 08.33'S	05:58:38	22.0	109	57	00m54.4s
017° 00.0'E	13° 08.09'S	13° 45.13'S	13° 26.52'S	05:59:08	22.6	109	58	00m55.4s
017° 30.0'E	13° 26.16'S	14° 03.77'S	13° 44.87'S	05:59:39	23.3	109	59	00m56.3s
018° 00.0'E	13° 44.39'S	14° 22.57'S	14° 03.38'S	06:00:11	24.0	108	59	00m57.3s
018° 30.0'E	14° 02.77'S	14° 41.52'S	14° 22.05'S	06:00:44	24.7	108	60	00m58.3s
019° 00.0'E	14° 21.31'S	15° 00.64'S	14° 40.88'S	06:01:18	25.4	108	61	00m59.3s
019° 30.0'E	14° 40.01'S	15° 19.91'S	14° 59.86'S	06:01:53	26.1	108	62	01m00.3s
020° 00.0'E	14° 58.85'S	15° 39.34'S	15° 19.00'S	06:02:30	26.8	108	62	01m01.3s
020° 30.0'E	15° 17.85'S	15° 58.91'S	15° 38.28'S	06:03:07	27.5	107	63	01m02.3s
021° 00.0'E	15° 37.00'S	16° 18.63'S	15° 57.72'S	06:03:46	28.2	107	64	01m03.3s
021° 30.0'E	15° 56.29'S	16° 38.50'S	16° 17.30'S	06:04:26	28.9	107	64	01m04.4s
022° 00.0'E	16° 15.73'S	16° 58.51'S	16° 37.02'S	06:05:06	29.6	107	65	01m05.4s
022° 30.0'E	16° 35.31'S	17° 18.66'S	16° 56.88'S	06:05:48	30.3	106	66	01m06.5s
023° 00.0'E	16° 55.02'S	17° 38.94'S	17° 16.87'S	06:06:32	31.0	106	66	01m07.5s
023° 30.0'E	17° 14.86'S	17° 59.35'S	17° 37.00'S	06:07:16	31.7	106	67	01m08.6s
024° 00.0'E	17° 34.84'S	18° 19.88'S	17° 57.25'S	06:08:01	32.5	105	68	01m09.7s
024° 30.0'E	17° 54.93'S	18° 40.54'S	18° 17.63'S	06:08:48	33.2	105	68	01m10.8s
025° 00.0'E	18° 15.15'S	19° 01.31'S	18° 38.12'S	06:09:36	33.9	105	69	01m11.9s
025° 30.0'E	18° 35.48'S	19° 22.19'S	18° 58.73'S	06:10:25	34.6	104	70	01m13.0s
026° 00.0'E	18° 55.92'S	19° 43.17'S	19° 19.44'S	06:11:15	35.4	104	70	01m14.1s
026° 30.0'E	19° 16.47'S	20° 04.25'S	19° 40.25'S	06:12:06	36.1	104	71	01m15.2s
027° 00.0'E	19° 37.12'S	20° 25.42'S	20° 01.16'S	06:12:58	36.9	103	72	01m16.3s
027° 30.0'E	19° 57.85'S	20° 46.67'S	20° 22.15'S	06:13:52	37.6	103	72	01m17.4s
028° 00.0'E	20° 18.68'S	21° 08.00'S	20° 43.23'S	06:14:47	38.3	102	73	01m18.5s
028° 30.0'E	20° 39.58'S	21° 29.40'S	21° 04.38'S	06:15:42	39.1	102	73	01m19.6s
029° 00.0'E	21° 00.55'S	21° 50.87'S	21° 25.60'S	06:16:39	39.8	102	74	01m20.8s
029° 30.0'E	21° 21.59'S	22° 12.38'S	21° 46.88'S	06:17:37	40.6	101	75	01m21.9s
030° 00.0'E	21° 42.69'S	22° 33.94'S	22° 08.21'S	06:18:37	41.3	101	75	01m23.0s
030° 30.0'E	22° 03.83'S	22° 55.54'S	22° 29.58'S	06:19:37	42.1	100	76	01m24.1s
031° 00.0'E	22° 25.02'S	23° 17.16'S	22° 50.99'S	06:20:38	42.8	100	76	01m25.3s
031° 30.0'E	22° 46.23'S	23° 38.80'S	23° 12.42'S	06:21:40	43.6	99	77	01m26.4s
032° 00.0'E	23° 07.47'S	24° 00.45'S	23° 33.86'S	06:22:44	44.3	99	77	01m27.5s
032° 30.0'E	23° 28.73'S	24° 22.10'S	23° 55.31'S	06:23:48	45.1	98	78	01m28.6s
033° 00.0'E	23° 49.98'S	24° 43.73'S	24° 16.76'S	06:24:53	45.8	98	78	01m29.7s
033° 30.0'E	24° 11.24'S	25° 05.35'S	24° 38.20'S	06:25:59	46.5	97	79	01m30.8s
034° 00.0'E	24° 32.48'S	25° 26.93'S	24° 59.61'S	06:27:06	47.3	96	79	01m31.9s
034° 30.0'E	24° 53.69'S	25° 48.47'S	25° 20.99'S	06:28:14	48.0	96	80	01m33.0s

- - - - - - - - - - - - - Begin 60' step interval in longitude - - - - - - - - - - - - -

035° 00.0'E	25° 14.87'S	26° 09.96'S	25° 42.33'S	06:29:23	48.8	95	80	01m34.1s
036° 00.0'E	25° 57.09'S	26° 52.73'S	26° 24.83'S	06:31:44	50.2	94	81	01m36.3s
037° 00.0'E	26° 39.05'S	27° 35.17'S	27° 07.04'S	06:34:07	51.7	92	82	01m38.4s
038° 00.0'E	27° 20.66'S	28° 17.19'S	27° 48.86'S	06:36:33	53.2	91	82	01m40.4s
039° 00.0'E	28° 01.86'S	28° 58.71'S	28° 30.23'S	06:39:01	54.6	89	83	01m42.4s

TABLE 8
MAPPING COORDINATES FOR THE UMBRAL PATH – AUSTRALIA
TOTAL SOLAR ECLIPSE OF 2002 DECEMBER 04

Longitude	Latitude of:			Circumstances on Center Line				
	Northern Limit	Southern Limit	Center Line	Universal Time h m s	Sun Alt °	Sun Az. °	Path Width km	Central Durat.
110°00.0'E	39°49.15'S	40°23.72'S	40°06.40'S	08:54:43	32.0	267	61	01m06.4s
111°00.0'E	39°33.77'S	40°07.95'S	39°50.83'S	08:55:43	31.0	266	60	01m04.9s
112°00.0'E	39°17.91'S	39°51.68'S	39°34.76'S	08:56:42	30.0	265	59	01m03.3s
113°00.0'E	39°01.57'S	39°34.92'S	39°18.20'S	08:57:39	29.1	264	58	01m01.8s
114°00.0'E	38°44.76'S	39°17.67'S	39°01.17'S	08:58:34	28.1	263	57	01m00.3s
115°00.0'E	38°27.48'S	38°59.95'S	38°43.68'S	08:59:28	27.1	262	56	00m58.8s
116°00.0'E	38°09.76'S	38°41.77'S	38°25.73'S	09:00:19	26.1	261	55	00m57.3s
117°00.0'E	37°51.61'S	38°23.14'S	38°07.33'S	09:01:10	25.1	261	54	00m55.8s
118°00.0'E	37°33.02'S	38°04.08'S	37°48.51'S	09:01:58	24.2	260	53	00m54.3s
119°00.0'E	37°14.02'S	37°44.58'S	37°29.26'S	09:02:44	23.2	259	52	00m52.8s
120°00.0'E	36°54.62'S	37°24.67'S	37°09.60'S	09:03:29	22.2	258	51	00m51.3s
121°00.0'E	36°34.83'S	37°04.35'S	36°49.55'S	09:04:12	21.2	257	50	00m49.9s
122°00.0'E	36°14.66'S	36°43.65'S	36°29.11'S	09:04:53	20.2	257	49	00m48.4s
123°00.0'E	35°54.12'S	36°22.56'S	36°08.30'S	09:05:32	19.2	256	48	00m47.0s
124°00.0'E	35°33.24'S	36°01.12'S	35°47.13'S	09:06:09	18.2	255	47	00m45.6s
125°00.0'E	35°12.01'S	35°39.32'S	35°25.62'S	09:06:45	17.2	255	46	00m44.2s
126°00.0'E	34°50.46'S	35°17.19'S	35°03.78'S	09:07:18	16.2	254	45	00m42.8s
127°00.0'E	34°28.60'S	34°54.73'S	34°41.62'S	09:07:50	15.2	253	44	00m41.4s
128°00.0'E	34°06.44'S	34°31.97'S	34°19.16'S	09:08:19	14.2	253	43	00m40.0s
129°00.0'E	33°43.99'S	34°08.91'S	33°56.41'S	09:08:47	13.2	252	42	00m38.7s
130°00.0'E	33°21.28'S	33°45.58'S	33°33.38'S	09:09:12	12.2	251	41	00m37.4s
131°00.0'E	32°58.31'S	33°21.98'S	33°10.10'S	09:09:36	11.3	251	40	00m36.0s
132°00.0'E	32°35.10'S	32°58.13'S	32°46.57'S	09:09:57	10.3	250	39	00m34.7s

- - - - - - - - - - - - - Begin 30' step interval in longitude - - - - - - - - - - - -

132°00.0'E	32°35.10'S	32°58.13'S	32°46.57'S	09:09:57	10.3	250	39	00m34.7s
132°30.0'E	32°23.41'S	32°46.11'S	32°34.72'S	09:10:07	9.8	250	38	00m34.1s
133°00.0'E	32°11.66'S	32°34.04'S	32°22.81'S	09:10:17	9.3	249	37	00m33.5s
133°30.0'E	31°59.86'S	32°21.92'S	32°10.85'S	09:10:26	8.8	249	37	00m32.8s
134°00.0'E	31°48.01'S	32°09.74'S	31°58.84'S	09:10:34	8.3	249	36	00m32.2s
134°30.0'E	31°36.11'S	31°57.51'S	31°46.77'S	09:10:42	7.8	249	36	00m31.6s
135°00.0'E	31°24.16'S	31°45.23'S	31°34.66'S	09:10:50	7.3	248	35	00m30.9s
135°30.0'E	31°12.17'S	31°32.90'S	31°22.50'S	09:10:57	6.8	248	35	00m30.3s
136°00.0'E	31°00.13'S	31°20.53'S	31°10.29'S	09:11:03	6.3	248	34	00m29.7s
136°30.0'E	30°48.05'S	31°08.11'S	30°58.04'S	09:11:09	5.8	248	33	00m29.1s
137°00.0'E	30°35.92'S	30°55.66'S	30°45.75'S	09:11:15	5.3	247	33	00m28.5s
137°30.0'E	30°23.76'S	30°43.16'S	30°33.42'S	09:11:20	4.8	247	32	00m27.9s
138°00.0'E	30°11.56'S	30°30.62'S	30°21.05'S	09:11:25	4.3	247	32	00m27.3s
138°30.0'E	29°59.33'S	30°18.04'S	30°08.65'S	09:11:29	3.8	246	31	00m26.7s
139°00.0'E	29°47.06'S	30°05.44'S	29°56.21'S	09:11:32	3.3	246	31	00m26.1s
139°30.0'E	29°34.75'S	29°52.79'S	29°43.74'S	09:11:35	2.9	246	30	00m25.6s
140°00.0'E	29°22.42'S	29°40.12'S	29°31.23'S	09:11:38	2.4	246	29	00m25.0s
140°30.0'E	29°10.06'S	29°27.41'S	29°18.70'S	09:11:40	1.9	245	29	00m24.4s
141°00.0'E	28°57.66'S	29°14.68'S	29°06.14'S	09:11:42	1.4	245	28	00m23.8s
141°30.0'E	28°45.25'S	29°01.92'S	28°53.55'S	09:11:43	0.9	245	28	00m23.3s

TABLE 9
MAPPING COORDINATES FOR THE ZONES OF GRAZING ECLIPSE – AFRICA
TOTAL SOLAR ECLIPSE OF 2002 DECEMBER 04

Longitude	North Graze Zone Latitudes		Northern Limit	South Graze Zone Latitudes		Southern Limit	Path Azm	Elev Fact	Scale Fact
	Northern Limit	Southern Limit	Universal Time	Northern Limit	Southern Limit	Universal Time			
° ′	° ′	° ′	h m s	° ′	° ′	h m s	°		km/"
013 00.0E	10 48.07S	10 49.19S	05:55:20	11 21.82S	11 24.29S	05:55:55	119.9	0.55	2.02
013 30.0E	11 04.86S	11 05.95S	05:55:42	11 39.14S	11 41.60S	05:56:19	120.2	0.55	2.02
014 00.0E	11 21.81S	11 22.88S	05:56:06	11 56.61S	11 59.07S	05:56:43	120.5	0.55	2.02
014 30.0E	11 38.93S	11 39.97S	05:56:30	12 14.26S	12 16.71S	05:57:08	120.7	0.55	2.02
015 00.0E	11 56.21S	11 57.23S	05:56:55	12 32.07S	12 34.51S	05:57:35	121.0	0.55	2.02
015 30.0E	12 13.66S	12 14.65S	05:57:21	12 50.05S	12 52.48S	05:58:02	121.3	0.55	2.02
016 00.0E	12 31.28S	12 32.23S	05:57:49	13 08.20S	13 10.61S	05:58:31	121.5	0.55	2.02
016 30.0E	12 49.06S	12 49.97S	05:58:17	13 26.50S	13 28.90S	05:59:00	121.8	0.55	2.02
017 00.0E	13 07.00S	13 07.88S	05:58:46	13 44.99S	13 47.38S	05:59:30	122.0	0.55	2.02
017 30.0E	13 25.11S	13 25.95S	05:59:16	14 03.54S	14 06.03S	06:00:02	122.3	0.55	2.02
018 00.0E	13 43.38S	13 44.17S	05:59:48	14 22.28S	14 24.86S	06:00:35	122.6	0.55	2.02
018 30.0E	14 01.80S	14 02.50S	06:00:20	14 41.25S	14 43.87S	06:01:08	122.8	0.55	2.02
019 00.0E	14 20.39S	14 21.04S	06:00:54	15 00.36S	15 03.03S	06:01:43	123.1	0.55	2.02
019 30.0E	14 39.13S	14 39.74S	06:01:28	15 19.63S	15 22.34S	06:02:19	123.3	0.55	2.02
020 00.0E	14 58.02S	14 58.58S	06:02:04	15 39.05S	15 41.80S	06:02:56	123.6	0.55	2.02
020 30.0E	15 17.07S	15 17.58S	06:02:41	15 58.62S	16 01.42S	06:03:34	123.8	0.55	2.02
021 00.0E	15 36.27S	15 36.73S	06:03:19	16 18.34S	16 21.17S	06:04:14	124.1	0.55	2.02
021 30.0E	15 55.61S	15 56.02S	06:03:58	16 38.21S	16 41.07S	06:04:54	124.3	0.55	2.02
022 00.0E	16 15.10S	16 15.46S	06:04:38	16 58.22S	17 01.11S	06:05:36	124.5	0.55	2.02
022 30.0E	16 34.67S	16 35.03S	06:05:19	17 18.36S	17 21.29S	06:06:18	124.8	0.54	2.01
023 00.0E	16 54.36S	16 54.74S	06:06:02	17 38.64S	17 41.59S	06:07:02	125.0	0.54	2.01
023 30.0E	17 14.17S	17 14.59S	06:06:45	17 59.05S	18 02.03S	06:07:47	125.2	0.54	2.01
024 00.0E	17 34.12S	17 34.56S	06:07:30	18 19.58S	18 22.58S	06:08:34	125.4	0.54	2.01
024 30.0E	17 54.20S	17 54.66S	06:08:16	18 40.23S	18 43.26S	06:09:21	125.7	0.54	2.01
025 00.0E	18 14.39S	18 14.88S	06:09:03	19 01.00S	19 04.05S	06:10:09	125.9	0.54	2.01
025 30.0E	18 34.71S	18 35.21S	06:09:51	19 21.88S	19 24.94S	06:10:59	126.1	0.53	2.01
026 00.0E	18 55.13S	18 55.65S	06:10:40	19 42.86S	19 45.94S	06:11:50	126.3	0.53	2.01
026 30.0E	19 15.66S	19 16.20S	06:11:31	20 03.94S	20 07.03S	06:12:42	126.4	0.53	2.00
027 00.0E	19 36.29S	19 36.84S	06:12:22	20 25.11S	20 28.21S	06:13:35	126.6	0.53	2.00
027 30.0E	19 57.01S	19 57.58S	06:13:15	20 46.36S	20 49.48S	06:14:29	126.8	0.53	2.00
028 00.0E	20 17.83S	20 18.40S	06:14:09	21 07.69S	21 10.82S	06:15:25	127.0	0.52	2.00
028 30.0E	20 38.72S	20 39.31S	06:15:04	21 29.09S	21 32.23S	06:16:21	127.1	0.52	2.00
029 00.0E	20 59.68S	21 00.28S	06:16:00	21 50.55S	21 53.69S	06:17:19	127.3	0.52	1.99
029 30.0E	21 20.72S	21 21.32S	06:16:58	22 12.07S	22 15.21S	06:18:18	127.4	0.52	1.99
030 00.0E	21 41.81S	21 42.42S	06:17:56	22 33.62S	22 36.78S	06:19:17	127.5	0.51	1.99
030 30.0E	22 02.94S	22 03.56S	06:18:56	22 55.22S	22 58.37S	06:20:18	127.7	0.51	1.99
031 00.0E	22 24.12S	22 24.74S	06:19:56	23 16.84S	23 20.00S	06:21:20	127.8	0.51	1.98
031 30.0E	22 45.34S	22 45.96S	06:20:58	23 38.48S	23 41.64S	06:22:23	127.9	0.50	1.98
032 00.0E	23 06.57S	23 07.20S	06:22:01	24 00.14S	24 03.29S	06:23:27	128.0	0.50	1.98
032 30.0E	23 27.83S	23 28.45S	06:23:04	24 21.84S	24 24.92S	06:24:32	128.0	0.50	1.98
033 00.0E	23 49.06S	23 49.72S	06:24:09	24 43.47S	24 46.55S	06:25:38	128.1	0.49	1.97
033 30.0E	24 10.31S	24 10.97S	06:25:15	25 05.09S	25 08.16S	06:26:45	128.2	0.49	1.97
034 00.0E	24 31.55S	24 32.21S	06:26:21	25 26.67S	25 29.74S	06:27:52	128.2	0.49	1.97
034 30.0E	24 52.77S	24 53.42S	06:27:29	25 48.21S	25 51.28S	06:29:01	128.2	0.48	1.97
035 00.0E	25 13.96S	25 14.60S	06:28:37	26 09.70S	26 12.76S	06:30:10	128.3	0.48	1.96
035 30.0E	25 35.10S	25 35.74S	06:29:46	26 31.13S	26 34.17S	06:31:20	128.3	0.48	1.96

Table 10

Mapping Coordinates for the Zones of Grazing Eclipse – Australia
Total Solar Eclipse of 2002 December 04

Longitude	North Graze Zone Latitudes		Northern Limit	South Graze Zone Latitudes		Southern Limit	Path Azm	Elev Fact	Scale Fact
	Northern Limit	Southern Limit	Universal Time	Northern Limit	Southern Limit	Universal Time			
° ´	° ´	° ´	h m s	° ´	° ´	h m s	°		km/"
130 00.0E	33 20.65S	33 21.57S	09:09:23	33 44.90S	33 47.91S	09:09:01	65.1	0.49	1.98
130 30.0E	33 09.23S	33 10.14S	09:09:35	33 33.09S	33 36.08S	09:09:13	65.1	0.50	1.98
131 00.0E	32 57.70S	32 58.59S	09:09:46	33 21.26S	33 24.24S	09:09:25	65.0	0.50	1.98
131 30.0E	32 46.13S	32 47.02S	09:09:57	33 09.38S	33 12.35S	09:09:36	64.9	0.50	1.98
132 00.0E	32 34.51S	32 35.39S	09:10:07	32 57.43S	33 00.39S	09:09:47	64.9	0.50	1.98
132 30.0E	32 22.83S	32 23.71S	09:10:17	32 45.41S	32 48.37S	09:09:57	64.8	0.50	1.98
133 00.0E	32 11.10S	32 11.97S	09:10:26	32 33.34S	32 36.28S	09:10:07	64.8	0.50	1.98
133 30.0E	31 59.31S	32 00.17S	09:10:35	32 21.21S	32 24.15S	09:10:16	64.7	0.51	1.98
134 00.0E	31 47.47S	31 48.33S	09:10:43	32 09.03S	32 11.96S	09:10:25	64.7	0.51	1.99
134 30.0E	31 35.58S	31 36.43S	09:10:51	31 56.80S	31 59.72S	09:10:33	64.6	0.51	1.99
135 00.0E	31 23.65S	31 24.49S	09:10:59	31 44.52S	31 47.43S	09:10:41	64.6	0.51	1.99
135 30.0E	31 11.66S	31 12.50S	09:11:05	31 32.19S	31 35.10S	09:10:48	64.6	0.51	1.99
136 00.0E	30 59.64S	31 00.46S	09:11:12	31 19.82S	31 22.71S	09:10:55	64.5	0.52	1.99
136 30.0E	30 47.57S	30 48.39S	09:11:17	31 07.40S	31 10.29S	09:11:01	64.5	0.52	2.00
137 00.0E	30 35.46S	30 36.27S	09:11:23	30 54.94S	30 57.82S	09:11:07	64.5	0.52	2.00
137 30.0E	30 23.31S	30 24.11S	09:11:28	30 42.44S	30 45.31S	09:11:12	64.4	0.53	2.00
138 00.0E	30 11.12S	30 11.92S	09:11:32	30 29.90S	30 32.77S	09:11:17	64.4	0.53	2.00
138 30.0E	29 58.90S	29 59.69S	09:11:36	30 17.32S	30 20.18S	09:11:21	64.4	0.53	2.00
139 00.0E	29 46.64S	29 47.43S	09:11:39	30 04.71S	30 07.56S	09:11:25	64.4	0.53	2.01
139 30.0E	29 34.31S	29 35.19S	09:11:42	29 52.07S	29 54.91S	09:11:29	64.4	0.53	2.01
140 00.0E	29 22.46S	29 22.46S	09:11:44	29 39.39S	29 42.23S	09:11:31	64.4	0.54	2.01
140 30.0E	29 10.09S	29 10.09S	09:11:46	29 26.67S	29 29.51S	09:11:34	64.4	0.55	2.02
141 00.0E	28 57.70S	28 57.70S	09:11:48	29 13.93S	29 16.76S	09:11:35	64.4	0.57	2.04
141 30.0E	28 45.26S	28 45.26S	09:11:49	29 01.12S	29 03.95S	09:11:37	64.4	0.60	2.07

TABLE 11
LOCAL CIRCUMSTANCES FOR AFRICA: ANGOLA – BENIN
TOTAL SOLAR ECLIPSE OF 2002 DECEMBER 04

Location Name	Latitude	Longitude	Elev. m	First Contact U.T. h m s	P °	V °	Alt °	Second Contact U.T. h m s	P °	V °	Third Contact U.T. h m s	P °	V °	Fourth Contact U.T. h m s	P °	V °	Alt °	Maximum Eclipse U.T. h m s	P °	V °	Alt °	Azm °	Eclip. Mag.	Eclip. Obs.	Umbral Depth	Umbral Durat.
ANGOLA																										
Alto-Uama	12°14'S	015°33'E	--	04:58:48.9	294	34	7	--	--	--	--	--	--	07:03:05.9	117	206	36	05:57:21.1	205	300	21	109	0.999	1.000		
Amboiva	11°32'S	014°44'E	--	04:58:17.2	293	33	6	--	--	--	--	--	--	07:01:34.4	117	206	34	05:56:23.5	205	300	19	110	0.993	0.994		
Bailundo	12°12'S	015°52'E	--	04:58:43.6	293	33	7	--	--	--	--	--	--	07:03:18.1	117	206	36	05:57:23.2	205	300	21	109	0.994	0.995		
Balombo	12°21'S	014°46'E	--	04:59:04.9	294	35	7	05:57:04.9	172	267	05:57:33.1	239	334	07:02:38.7	116	206	35	05:57:19.0	25	121	20	109	1.001	1.000	0.166	00m28s
Bando	15°00'S	020°30'E	--	04:59:36.4	294	35	13	--	--	--	--	--	--	07:11:23.2	118	208	43	06:02:17.7	206	302	27	108	0.993	0.994		
Bela Vista	12°35'S	016°13'E	--	04:59:03.1	294	34	8	--	--	--	--	--	--	07:04:05.4	117	206	37	05:57:54.8	205	301	21	109	0.998	0.999		
Benguela	12°35'S	013°25'E	--	04:59:37.8	295	37	6	--	--	--	--	--	--	07:01:55.6	114	206	34	05:56:24.0	205	300	19	109	0.976	0.974		
Botera	11°37'S	014°17'E	--	04:58:27.6	294	34	6	05:56:07.4	66	162	05:56:40.7	343	78	07:01:21.2	116	206	34	05:56:24.0	25	121	19	109	1.000	1.000	0.252	00m33s
Cadia	12°51'S	015°33'E	--	04:59:26.3	294	35	8	--	--	--	--	--	--	07:03:54.2	116	206	36	05:58:03.7	205	301	21	109	1.000	1.000		
Cachingues	13°05'S	016°43'E	--	04:59:29.1	294	35	9	05:58:16.2	78	173	05:58:59.4	333	68	07:05:10.3	117	206	37	05:58:37.7	205	301	22	109	1.003	1.000	0.388	00m43s
Capuna	15°38'S	019°43'E	--	05:01:58.1	295	37	13	--	--	--	--	--	--	07:11:30.1	116	207	43	06:02:47.5	26	123	27	107	0.996	0.997		
Cassongue	11°51'S	015°03'E	--	04:58:31.8	293	34	7	--	--	--	--	--	--	07:02:13.0	117	206	35	05:56:48.5	205	300	20	110	0.997	0.998		
Catota	13°52'S	017°15'E	--	05:00:14.2	295	36	10	05:59:29.4	176	272	05:59:56.9	235	331	07:06:41.4	118	207	40	05:59:43.1	26	121	24	109	0.990	0.991	0.129	00m28s
Chá Pungana	13°44'S	018°39'E	--	05:00:00.2	294	34	11	--	--	--	--	--	--	07:07:46.2	118	207	40	06:00:00.2	206	301	24	108	0.990	0.991		
Chila	12°04'S	014°29'E	--	04:58:51.5	294	35	6	05:56:33.5	141	236	05:57:19.3	270	5	07:02:04.3	116	205	34	05:56:56.4	25	120	19	110	1.004	1.000	0.570	00m46s
Chinguar	12°36'S	016°20'E	--	04:59:03.0	294	34	8	--	--	--	--	--	--	07:04:12.4	117	206	37	05:57:57.7	205	301	22	109	0.997	0.998		
Chissilo	13°34'S	016°30'E	--	05:00:01.5	295	36	9	--	--	--	--	--	--	07:05:38.0	116	206	38	05:59:08.7	25	121	22	109	0.997	0.998		
Chitembo	13°34'S	016°40'E	--	05:00:00.0	295	36	9	--	--	--	--	--	--	07:05:46.6	116	206	38	05:59:11.4	25	121	22	109	1.000	1.000		
Chiumbe	12°29'S	016°08'E	--	04:58:57.9	294	34	8	--	--	--	--	--	--	07:03:53.4	116	206	37	05:57:46.6	205	301	21	109	0.997	0.998		
Ciuma	13°14'S	015°40'E	--	04:59:48.9	295	36	8	--	--	--	--	--	--	07:04:29.9	115	206	36	05:58:32.2	25	121	21	109	0.993	0.994		
Covelo	12°06'S	013°55'E	--	04:59:01.0	294	35	6	06:04:33.9	89	186	06:05:32.5	323	60	07:01:41.7	115	205	34	05:56:52.4	25	121	19	110	0.995	0.996		
Cuando	16°32'S	022°07'E	--	05:02:58.7	295	37	16	06:00:39.1	84	180	06:01:29.5	327	63	07:15:23.9	117	208	46	05:56:03.1	206	303	30	107	1.005	1.000	0.550	00m59s
Cuango	14°30'S	018°59'E	--	05:00:44.4	294	35	12	--	--	--	--	--	--	07:09:09.9	117	207	41	06:01:04.2	206	302	25	108	1.004	1.000	0.477	00m50s
Cuito-Cuanavale	15°10'S	019°10'E	--	05:01:28.3	295	37	12	--	--	--	--	--	--	07:10:17.1	116	207	41	06:01:58.8	25	120	19	108	0.999	0.999		
Cuma	12°52'S	015°05'E	--	04:59:32.7	295	36	7	--	--	--	--	--	--	07:03:33.2	115	205	35	05:57:58.5	25	122	20	109	0.993	0.994		
Dima	15°27'S	020°10'E	--	05:01:44.0	295	36	13	06:02:13.6	120	217	06:03:15.0	291	28	07:11:42.0	117	207	42	06:02:44.2	26	122	27	107	1.007	1.000	0.922	01m01s
Dumbo	14°06'S	017°24'E	--	05:00:28.2	295	36	10	--	--	--	--	--	--	07:07:09.0	116	206	39	06:00:02.9	25	121	23	108	0.998	0.999		
Ebanga	12°44'S	014°44'E	--	04:59:28.7	295	36	7	05:56:10.2	127	222	05:56:59.1	283	18	07:03:06.6	115	206	35	05:57:44.8	25	121	20	109	0.991	0.992	0.791	00m49s
Gungo	11°48'S	014°08'E	--	04:58:40.3	294	34	6	05:57:32.5	120	215	05:58:25.4	291	26	07:03:53.4	116	206	36	05:56:34.6	25	120	19	110	1.005	1.000	0.919	00m53s
Huambo	12°44'S	015°47'E	--	04:59:16.6	294	35	8	--	--	--	--	--	--	07:04:29.9	115	206	36	05:57:58.9	25	121	21	109	1.007	1.000		
Lépi	12°52'S	015°26'E	--	04:59:28.6	295	36	7	--	--	--	--	--	--	07:03:49.9	116	206	36	05:58:03.2	25	121	21	109	0.998	0.999		
Liutunga	13°17'S	016°43'E	--	04:59:41.6	294	35	9	--	--	--	--	--	--	07:05:26.4	116	206	38	05:58:52.0	25	121	22	109	1.005	1.000		
Lobito	12°20'S	013°34'E	--	04:59:20.1	295	37	6	05:58:24.6	118	214	05:59:19.4	293	28	07:01:43.8	115	205	34	05:57:07.4	25	121	19	110	0.984	0.984	0.956	00m55s
Longa	14°42'S	018°32'E	--	05:01:00.1	295	36	11	--	--	--	--	--	--	07:09:01.0	116	206	41	06:01:09.6	26	122	25	108	1.000	1.000		
Luanda	08°48'S	013°14'E	59	04:56:11.5	293	28	3	--	--	--	--	--	--	06:57:06.1	119	206	31	05:53:15.6	205	298	16	111	0.946	0.936		
Luatira	12°52'S	017°14'E	--	04:59:32.7	295	34	9	--	--	--	--	--	--	07:05:18.9	117	206	38	05:58:31.0	205	301	23	109	0.990	0.991		
Lubango	14°55'S	013°30'E	--	05:02:10.0	299	42	7	--	--	--	--	--	--	07:04:49.9	111	204	35	05:58:58.0	205	301	23	109	0.897	0.873		
Luiana	17°23'S	023°03'E	--	05:02:28.2	295	37	17	06:06:08.5	127	224	06:07:15.0	285	23	07:17:45.9	117	209	48	06:06:41.7	26	123	29	106	0.922	0.905	0.813	01m07s
Luimbale	12°15'S	015°19'E	--	04:58:52.4	294	34	7	05:56:58.8	76	172	05:57:39.5	334	69	07:02:56.3	116	206	35	05:57:19.1	205	300	21	109	1.007	1.000	0.376	00m41s
Lunge	12°12'S	016°05'E	--	04:59:01.4	293	34	8	--	--	--	--	--	--	07:03:28.4	117	206	36	05:57:26.3	205	301	21	109	1.003	1.000		
Lupire	14°36'S	019°29'E	--	05:00:00.1	294	35	12	--	--	--	--	--	--	07:09:47.5	117	207	42	06:01:22.8	206	302	26	108	0.999	0.999		
Mavinga	15°50'S	020°21'E	--	05:02:10.1	295	37	14	06:03:02.6	175	271	06:03:35.0	237	334	07:12:26.7	117	206	43	06:03:18.8	26	123	27	107	1.001	1.000	0.147	00m32s
Menongue	14°36'S	017°48'E	--	05:00:58.2	295	37	11	--	--	--	--	--	--	07:08:11.2	116	206	40	06:00:47.3	206	303	24	108	0.992	0.993		
Missao Santa Cr...	16°14'S	021°57'E	--	05:02:58.7	295	36	15	06:04:29.4	34	131	06:04:38.8	18	114	07:14:45.7	117	208	45	06:04:34.0	206	303	29	107	1.000	1.000	0.010	00m09s
Moçâmedes	15°10'S	012°09'E	--	05:02:50.6	300	44	6	--	--	--	--	--	--	07:04:02.4	109	203	34	06:00:06.2	25	124	19	109	1.000	1.000		
Mutumbo	13°14'S	017°17'E	--	04:59:33.8	294	33	9	--	--	--	--	--	--	07:05:55.1	117	206	38	05:58:58.0	205	301	23	109	0.998	0.999		
Ngunza	11°13'S	013°50'E	--	04:58:10.7	293	33	5	--	--	--	--	--	--	07:00:31.7	115	205	33	05:55:53.0	205	300	18	110	0.998	0.999		
Quinjenje	12°49'S	014°55'E	--	04:59:31.6	295	36	7	--	--	--	--	--	--	07:03:21.5	115	206	35	05:57:34.1	205	301	20	109	1.000	1.000		
Quipeio	12°26'S	015°30'E	--	04:59:01.6	294	34	7	05:57:09.8	94	189	05:57:58.6	317	52	07:03:19.2	116	206	36	05:57:34.1	205	301	20	109	1.005	1.000	0.635	00m49s
Samacimbo	13°33'S	016°59'E	--	04:59:56.1	294	35	9	05:58:50.0	138	234	05:59:41.2	273	9	07:06:01.7	116	206	38	05:59:15.5	25	121	23	109	1.005	1.000	0.618	00m51s
Sambo	12°57'S	016°05'E	--	04:59:26.8	294	35	8	05:57:52.2	127	223	05:58:44.7	283	19	07:04:28.1	116	206	37	05:58:18.4	25	121	23	109	1.006	1.000	0.795	00m53s
Sandumba	13°45'S	017°29'E	--	05:00:00.9	294	34	10	05:59:10.8	118	214	06:00:07.0	293	28	07:06:44.3	116	206	39	05:59:38.9	25	121	23	109	1.007	1.000	0.961	00m56s
Saupite	13°54'S	017°43'E	--	05:00:12.1	294	35	10	05:59:25.8	119	215	06:00:22.5	292	28	07:07:09.1	116	206	39	05:59:54.1	25	121	24	108	1.007	1.000	0.941	00m57s
Vila Nova	12°38'S	016°03'E	--	04:59:07.8	294	34	8	05:57:39.4	63	159	05:58:12.3	347	82	07:04:01.3	117	206	37	05:57:55.8	205	301	21	109	1.002	1.000	0.213	00m33s
BENIN																										
Abomey	07°11'N	001°59'E	--	--	--	--	--	--	--	--	--	--	--	06:36:30.3	134	212	10	05:52 Rise			0	112	0.623	0.532		
Cotonou	06°21'N	002°26'E	--	--	--	--	--	--	--	--	--	--	--	06:37:20.7	133	212	11	05:48 Rise			0	112	0.661	0.579		
Parakou	09°21'N	002°37'E	--	--	--	--	--	--	--	--	--	--	--	06:33:54.1	138	214	9	05:53 Rise			0	112	0.544	0.441		
Porto-Novo	06°29'N	002°37'E	--	--	--	--	--	--	--	--	--	--	--	06:37:10.2	133	212	11	05:48 Rise			0	112	0.658	0.574		

TABLE 12
LOCAL CIRCUMSTANCES FOR AFRICA: BOTSWANA – EQUATORIAL GUINEA
TOTAL SOLAR ECLIPSE OF 2002 DECEMBER 04

Location Name	Latitude	Longitude	Elev.	First Contact U.T. h m s	P °	V °	Alt °	Second Contact U.T. h m s	P °	V °	Third Contact U.T. h m s	P °	V °	Fourth Contact U.T. h m s	P °	V °	Alt °	Maximum Eclipse U.T. h m s	P °	V °	Alt °	Azm °	Eclip. Mag.	Eclip. Obs.	Umbral Depth	Umbral Durat.
BOTSWANA			m																							
Chobe	17°52'S	025°07'E	—	05:04:49.8	294	36	19	—			—			07:21:03.2	118	210	50	06:08:33.9	206	304	34	105	0.989	0.990		
Dodo	18°45'S	025°20'E	—	05:05:58.8	295	38	20	06:09:23.7	99	197	06:10:33.0	314	51	07:22:44.2	117	210	51	06:09:58.3	206	304	34	105	1.006	1.000	0.705	01m09s
Francistown	21°11'S	027°32'E	1004	05:09:38.0	297	41	23	—			—			07:29:39.1	116	211	55	06:15:07.2	27	126	38	102	0.992	0.992		
Gaborone	24°45'S	025°55'E	—	05:14:48.3	302	50	24	—			—			07:32:41.0	110	211	55	06:19:30.3	26	131	38	100	0.886	0.860		
Gerufa	19°17'S	026°02'E	—	05:06:46.7	295	38	21	06:10:36.1	107	206	06:11:49.2	306	44	07:24:30.8	117	210	52	06:09:11.2.5	206	305	35	104	1.008	1.000	0.838	01m13s
Kakoaka	18°40'S	024°22'E	—	05:05:45.8	296	39	19	—			—			07:21:21.5	116	209	50	06:09:15.3	26	125	33	105	0.998	0.998		
Kalakamate	20°39'S	027°21'E	—	05:08:51.7	296	40	23	06:13:58.6	188	287	06:14:23.3	225	325	07:28:32.4	117	211	55	06:14:10.9	27	126	38	103	1.000	1.000	0.053	01m25s
Kasinka	18°13'S	024°22'E	—	05:05:11.1	295	38	19	06:08:01.5	118	216	06:09:11.9	294	32	07:20:39.5	117	209	50	06:08:36.6	26	124	33	105	1.000	1.000	0.964	01m10s
Kavimba	18°02'S	024°38'E	—	05:04:58.8	295	37	19	06:08:17.2	49	146	06:08:44.0	4	101	07:20:42.4	117	209	50	06:08:30.5	206	304	33	105	1.001	1.000	0.074	00m27s
Kumba Pits	18°45'S	024°45'E	—	05:05:54.6	296	39	19	06:09:12.6	165	263	06:10:00.3	248	346	07:21:58.7	117	209	50	06:09:36.4	26	125	34	105	1.002	1.000	0.255	00m48s
Maitengwe	20°06'S	027°13'E	—	05:08:04.9	296	39	22	06:12:37.4	106	205	06:13:52.9	307	45	07:27:27.7	117	211	54	06:13:15.1	207	305	37	103	1.008	1.000	0.825	01m16s
Maun	20°00'S	023°24'E	—	05:07:29.1	298	43	19	—			—			07:22:08.3	114	209	49	06:10:36.6	26	126	33	104	0.952	0.944		
Matlamanyane	19°33'S	025°57'E	—	05:07:07.2	296	39	21	06:11:05.0	158	257	06:12:00.7	255	354	07:24:49.5	117	210	52	06:11:32.8	26	125	35	104	1.003	1.000	0.339	00m56s
Nata	20°12'S	026°12'E	—	05:08:02.2	296	40	21	—			—			07:26:11.9	116	210	52	06:11:41.0	26	126	36	103	0.993	0.993		
Old Tate	21°22'S	027°46'E	—	05:09:57.3	297	41	24	—			—			07:30:18.0	116	212	56	06:15:35.5	27	127	38	102	0.991	0.992		
Pandamatenga	18°35'S	025°42'E	—	05:05:49.0	295	37	20	—			—			07:22:57.0	118	210	51	06:09:57.3	206	304	35	105	0.997	0.998		
Ramokawebana	20°33'S	027°48'E	—	05:08:48.2	296	41	23	06:13:43.1	112	212	06:15:00.9	301	40	07:29:01.6	117	211	55	06:14:27.9	207	306	38	103	1.009	1.000	0.927	01m18s
Serowe	22°24'S	026°42'E	—	05:11:15.6	299	44	23	—			—			07:30:21.0	114	211	55	06:16:23.5	26	128	38	101	0.950	0.942		
Tsekanyani	19°52'S	026°39'E	—	05:07:39.5	296	39	22	06:11:53.1	130	229	06:13:06.7	283	22	07:26:17.5	117	211	53	06:12:29.8	27	125	36	103	1.007	1.000	0.771	01m14s
BURKINA FASO																										
Bobo Dioulasso	11°12'N	004°18'W	—	—				—			—			06:34:04.1	136	213	2	06:23 Rise	—	—	0	113	0.171	0.083		
Koudougou	12°15'N	002°22'W	—	—				—			—			06:32:13.3	139	215	3	06:17 Rise	—	—	0	113	0.226	0.124		
Ouagadougou	12°22'N	001°31'W	—	—				—			—			06:31:45.5	140	216	4	06:14 Rise	—	—	0	113	0.261	0.154		
BURUNDI																										
Bujumbura	03°23'S	029°22'E	—	04:55:40.3	270	357	16	—			—			06:58:03.5	145	217	44	05:53:07.3	207	288	29	114	0.559	0.460		
CAMEROON																										
Douala	04°03'N	009°42'E	—	—				—			—			06:39:37.3	136	213	19	05:43:10.2	204	287	6	113	0.652	0.567		
Garoua	09°18'N	013°24'E	—	—				—			—			06:31:32.6	148	220	18	05:40:22.0	204	282	7	114	0.455	0.343		
Maroua	10°36'N	014°20'E	—	—				—			—			06:29:03.9	152	222	18	05:39:50.1	204	280	7	114	0.406	0.291		
Yaoundé	03°52'N	011°31'E	770	04:54:35.3	257	336	1	—			—			06:39:54.9	137	214	21	05:43:07.0	204	287	8	113	0.633	0.545		
CENTRAL AFRICAN REPUBLIC																										
Bangui	04°22'N	018°35'E	387	04:52:02.4	266	350	2	—			—			06:39:32.5	145	218	27	05:43:03.9	205	285	14	114	0.521	0.416		
CHAD																										
Moundou	08°34'N	016°05'E	—	—				—			—			06:32:16.4	150	221	21	05:40:38.6	205	282	9	114	0.440	0.327		
Ndjamena	12°07'N	015°03'E	295	—				—			—			06:25:59.7	156	225	17	05:39:18.2	204	279	7	114	0.353	0.238		
Sarh	09°09'N	018°23'E	—	04:54:35.3	257	336	1	—			—			06:30:41.7	154	223	22	05:40:30.7	205	280	11	115	0.391	0.276		
CISKEI																										
Potsdam	32°56'S	027°42'E	—	05:29:49.1	310	66	30	—			—			07:46:34.9	100	217	59	06:34:27.8	25	140	44	90	0.738	0.675		
CONGO																										
Brazzaville	04°16'S	015°17'E	318	04:52:55.6	282	16	3	—			—			06:52:08.8	128	210	30	05:49:11.7	205	293	16	112	0.801	0.752		
Dolisie	04°12'S	012°41'E	—	04:53:17.3	284	18	1	—			—			06:50:57.6	126	208	27	05:48:55.1	205	294	13	112	0.836	0.795		
Pointe-Noire	04°48'S	011°51'E	50	04:53:47.1	285	20	0	—			—			06:51:23.3	124	208	27	05:49:24.0	205	295	13	112	0.863	0.829		
DEMOCRATIC REPUBLIC OF THE CONGO																										
Kananga	05°54'S	022°25'E	—	04:53:54.1	280	12	10	—			—			06:58:36.5	133	212	39	05:52:28.1	206	293	24	112	0.737	0.673		
Kinshasa	04°18'S	015°18'E	—	04:52:56.8	282	16	3	—			—			06:52:12.4	128	210	30	05:49:13.9	205	293	16	112	0.802	0.752		
Kisangani	00°30'S	025°12'E	—	04:53:23.0	269	355	10	—			—			06:50:04.3	144	217	37	05:48:22.6	207	287	23	114	0.552	0.451		
Kolwezi	10°43'S	025°28'E	—	04:57:34.8	284	20	15	—			—			07:09:25.2	129	211	46	05:59:15.6	207	297	30	110	0.811	0.765		
Lubumbashi	11°40'S	027°28'E	—	04:58:53.3	284	20	18	—			—			07:13:07.7	130	212	49	06:01:36.2	207	293	32	109	0.802	0.754		
Mbuji-Mayi	06°09'S	023°38'E	—	04:54:15.3	279	11	11	—			—			06:59:51.2	134	213	40	05:53:11.7	206	293	25	112	0.725	0.657		
EQUATORIAL GUINEA																										
Malabo	03°45'N	008°47'E	—	—				—			—			06:39:59.5	134	212	18	05:43:26.4	204	288	5	113	0.672	0.591		

TABLE 13
LOCAL CIRCUMSTANCES FOR AFRICA: ETHIOPIA – MALI
TOTAL SOLAR ECLIPSE OF 2002 DECEMBER 04

Location Name	Latitude	Longitude	Elev. (m)	First Contact U.T. h m s	P °	V °	Alt °	Second Contact U.T. h m s	P °	V °	Third Contact U.T. h m s	P °	V °	Fourth Contact U.T. h m s	P °	V °	Alt °	Maximum Eclipse U.T. h m s	P °	V °	Alt °	Azm °	Eclip. Mag.	Eclip. Obs.	Umbral Depth	Umbral Durat.
ETHIOPIA																										
Adis Abeba	09°02'N	038°42'E	2450	05:23:57.3	229	296	25	– – –	–	–	– – –	–	–	06:15:18.9	188	247	36	05:49:03.4	208	272	31	123	0.066	0.020		
Akaki Beseka	08°52'N	038°47'E	–	05:23:37.8	229	297	25	– – –	–	–	– – –	–	–	06:16:09.4	187	247	36	05:49:17.0	208	272	31	123	0.069	0.022		
Bahir Dar	11°35'N	037°28'E	–	05:33:03.4	219	283	25	– – –	–	–	– – –	–	–	06:00:30.5	197	257	31	05:46:42.2	208	270	28	123	0.019	0.003		
Debre Zeyit	08°45'N	038°59'E	–	05:23:53.8	229	297	26	– – –	–	–	– – –	–	–	06:16:22.4	187	247	37	05:49:31.5	208	272	31	123	0.068	0.021		
Jima	07°36'N	036°50'E	–	05:13:54.5	238	309	22	– – –	–	–	– – –	–	–	06:25:43.5	178	239	37	05:48:34.6	208	274	30	121	0.138	0.060		
Nazret	08°33'N	039°16'E	–	05:24:08.4	229	297	26	– – –	–	–	– – –	–	–	06:16:53.3	187	247	37	05:49:53.8	208	272	32	123	0.069	0.021		
GABON																										
Franceville	01°38'S	013°35'E	–	04:52:04.8	279	11	0	– – –	–	–	– – –	–	–	06:47:51.7	130	211	27	05:46:51.4	205	291	13	112	0.755	0.694		
Libreville	00°23'N	009°27'E	35	– – –	–	–	–	– – –	–	–	– – –	–	–	06:44:16.1	130	210	21	05:45:27.2	204	291	8	112	0.756	0.695		
Port-Gentil	00°43'S	008°47'E	–	– – –	–	–	–	– – –	–	–	– – –	–	–	06:45:29.1	127	209	21	05:46:16.1	204	292	8	112	0.794	0.742		
GHANA																										
Accra	05°33'N	000°13'W	27	– – –	–	–	–	– – –	–	–	– – –	–	–	06:38:35.0	129	210	9	05:57 Rise	–	–	0	112	0.644	0.558		
Kumasi	06°41'N	001°35'W	287	– – –	–	–	–	– – –	–	–	– – –	–	–	06:37:43.1	130	210	7	06:03 Rise	–	–	0	112	0.559	0.458		
Tamale	09°25'N	000°50'W	–	– – –	–	–	–	– – –	–	–	– – –	–	–	06:34:45.6	136	213	6	06:07 Rise	–	–	0	112	0.429	0.315		
Tema	05°38'N	000°01'E	–	– – –	–	–	–	– – –	–	–	– – –	–	–	06:38:27.9	130	210	9	05:57 Rise	–	–	0	112	0.639	0.552		
GUINEA																										
Nzérékoré	07°45'N	008°49'W	–	– – –	–	–	–	– – –	–	–	– – –	–	–	06:38:44.9	127	208	1	06:36 Rise	–	–	0	112	0.054	0.015		
IVORY COAST																										
Abidjan	05°19'N	004°02'W	20	– – –	–	–	–	– – –	–	–	– – –	–	–	06:39:29.4	126	208	6	06:12 Rise	–	–	0	112	0.468	0.357		
Bouaké	07°41'N	005°02'W	364	– – –	–	–	–	– – –	–	–	– – –	–	–	06:37:38.1	129	210	4	06:18 Rise	–	–	0	112	0.337	0.223		
Daloa	06°53'N	006°27'W	–	– – –	–	–	–	– – –	–	–	– – –	–	–	06:38:43.9	127	208	3	06:25 Rise	–	–	0	112	0.240	0.137		
Korhogo	09°27'N	005°38'W	–	– – –	–	–	–	– – –	–	–	– – –	–	–	06:36:13.5	132	211	2	06:26 Rise	–	–	0	112	0.174	0.085		
Yamoussoukro	06°49'N	005°17'W	–	– – –	–	–	–	– – –	–	–	– – –	–	–	06:38:28.4	128	209	4	06:20 Rise	–	–	0	112	0.314	0.202		
KENYA																										
Eldoret	00°31'N	035°17'E	–	05:01:48.0	257	337	21	– – –	–	–	– – –	–	–	06:50:36.6	159	224	45	05:53:12.7	208	282	33	117	0.354	0.240		
Kisumu	00°06'S	034°45'E	–	05:00:49.5	259	340	21	– – –	–	–	– – –	–	–	06:52:15.4	157	223	45	05:53:24.1	208	283	33	117	0.380	0.266		
Machakos	01°31'S	037°16'E	–	05:03:21.2	258	340	24	– – –	–	–	– – –	–	–	06:57:03.7	158	222	49	05:56:35.5	208	283	36	118	0.372	0.258		
Meru	00°03'N	037°39'E	–	05:04:46.5	255	334	24	– – –	–	–	– – –	–	–	06:55:38.1	162	225	48	05:55:38.1	208	281	36	117	0.324	0.211		
Mombasa	04°03'S	039°40'E	16	05:05:49.6	260	343	28	– – –	–	–	– – –	–	–	07:05:44.5	156	220	55	06:02:08.4	209	284	41	116	0.394	0.280		
Nairobi	01°17'S	036°49'E	1820	05:02:49.2	259	340	24	– – –	–	–	– – –	–	–	06:56:12.0	158	222	49	05:56:14.8	208	283	36	117	0.374	0.260		
Nakuru	00°17'S	036°04'E	–	05:02:21.6	257	338	22	– – –	–	–	– – –	–	–	06:53:05.5	159	223	47	05:54:37.2	208	282	34	117	0.361	0.247		
LESOTHO																										
Maseru	29°28'S	027°30'E	–	05:23:09.4	307	59	28	– – –	–	–	– – –	–	–	07:41:49.4	105	215	58	06:28:26.6	26	136	42	94	0.808	0.762		
MADAGASCAR																										
Antananarivo	18°55'S	047°31'E	–	05:22:28.3	276	11	44	– – –	–	–	– – –	–	–	07:58:37.3	139	211	80	06:35:20.6	208	297	61	102	0.631	0.545		
Antsirabe	19°51'S	047°02'E	–	05:22:37.8	278	14	44	– – –	–	–	– – –	–	–	08:00:06.1	137	215	80	06:36:06.8	208	299	61	100	0.663	0.584		
Antsiranana	12°16'S	049°17'E	–	05:22:05.9	263	349	44	– – –	–	–	– – –	–	–	07:41:34.2	153	209	81	06:27:21.3	209	285	59	113	0.427	0.314		
Fianarantsoa	21°26'S	047°05'E	–	05:24:17.3	280	18	44	– – –	–	–	– – –	–	–	08:04:03.4	134	220	81	06:38:55.8	207	301	62	97	0.703	0.632		
Mahajanga	15°43'S	046°19'E	–	05:18:13.6	272	4	41	– – –	–	–	– – –	–	–	07:47:47.8	144	210	75	06:27:56.1	208	293	57	107	0.570	0.473		
Toamasina	18°10'S	049°23'E	–	05:24:52.8	273	6	46	– – –	–	–	– – –	–	–	08:00:04.0	142	205	81	06:37:24.4	208	294	63	103	0.578	0.483		
Toliara	23°21'S	043°40'E	–	05:22:23.8	286	28	41	– – –	–	–	– – –	–	–	08:01:32.4	128	225	78	06:36:42.4	207	306	58	95	0.810	0.765		
MALAWI																										
Blantyre	15°47'S	035°00'E	–	05:06:14.2	283	20	28	– – –	–	–	– – –	–	–	07:30:34.7	132	214	61	06:13:24.6	208	299	43	106	0.775	0.720		
Lilongwe	13°59'S	033°44'E	–	05:03:45.2	282	17	26	– – –	–	–	– – –	–	–	07:25:02.7	133	213	58	06:09:32.6	208	297	41	107	0.753	0.693		
Mzuzu	11°27'S	033°55'E	–	05:01:53.5	278	11	24	– – –	–	–	– – –	–	–	07:19:46.9	137	214	56	06:06:08.7	208	294	39	109	0.687	0.611		
MALI																										
Gao	16°16'N	000°03'W	–	– – –	–	–	–	– – –	–	–	– – –	–	–	06:26:14.3	149	221	2	06:15 Rise	–	–	0	113	0.150	0.068		
Mopti	14°30'N	004°12'W	–	– – –	–	–	–	– – –	–	–	– – –	–	–	06:30:31.3	142	217	0	06:29 Rise	–	–	0	113	0.029	0.006		
Sikasso	11°19'N	005°40'W	–	– – –	–	–	–	– – –	–	–	– – –	–	–	06:34:28.6	135	213	1	06:29 Rise	–	–	0	113	0.089	0.031		

TABLE 14
LOCAL CIRCUMSTANCES FOR AFRICA: MAYOTTE – SOMALIA
TOTAL SOLAR ECLIPSE OF 2002 DECEMBER 04

Location Name	Latitude	Longitude	Elev.	First Contact U.T. h m s	P °	V °	Alt °	Second Contact U.T. h m s	P °	V °	Third Contact U.T. h m s	P °	V °	Fourth Contact U.T. h m s	P °	V °	Alt °	Maximum Eclipse U.T. h m s	P °	V °	Alt °	Azm °	Eclip. Mag.	Eclip. Obs.	Umbral Depth	Umbral Durat.
MAYOTTE			m																							
Dzaoudzi	12°47'S	045°17'E	—	05:15:04.1	269	358	38	—			—			07:37:42.5	148	211	71	06:21:34.1	209	290	54	110	0.514	0.409		
MOZAMBIQUE																										
Beira	19°49'S	034°52'E	9	05:10:39.3	289	30	30	06:25:27.1	125	227	06:26:57.2	289	32	07:38:33.3	125	215	64	06:19:32.5	207	303	46	102	0.875	0.846	0.865	01m30s
Chibuto	24°44'S	033°33'E	—	05:16:52.1	296	42	31	06:26:32.3	84	187	06:27:49.6	330	73	07:45:12.1	117	218	65	06:26:12.1	27	130	47	97	1.009	1.000	0.454	01m17s
Chidenguele	24°54'S	034°10'E	—	05:17:26.3	296	42	32	06:26:10.7	138	241	06:27:36.2	276	19	07:46:03.5	117	218	65	06:27:10.9	27	130	48	96	1.005	1.000	0.645	01m26s
Chongoene	25°00'S	033°47'E	—	05:17:23.4	296	43	32	06:24:33.4	144	246	06:25:53.4	270	12	07:46:49.5	117	217	65	06:26:53.4	27	130	47	96	1.007	1.000	0.545	01m20s
Guija	24°29'S	033°00'E	—	05:16:14.5	296	42	30	—			—			07:47:06.0	120	217	67	06:25:13.3	27	129	46	97	1.006	0.954		
Inhambane	23°51'S	035°29'E	—	05:16:30.6	293	38	33	—			—			07:41:32.5	118	216	63	06:26:48.5	207	308	48	97	0.959	0.954	0.285	01m02s
Mabalane	23°37'S	032°31'E	—	05:14:43.2	296	41	30	06:22:47.9	71	173	06:23:49.8	342	84	07:45:03.3	116	218	63	06:23:18.8	207	308	45	98	1.003	1.000		
Macia	25°03'S	033°10'E	—	05:17:11.7	297	43	31	—			—			07:46:20.0	117	217	65	06:26:20.0	207	310	46	97	0.995	0.997		
Manjacaze	24°44'S	033°53'E	—	05:17:01.3	296	42	31	06:25:50.9	94	197	06:27:15.2	320	63	07:45:47.3	117	218	65	06:26:33.0	207	310	47	97	1.006	1.000	0.609	01m24s
Mapulanguene	24°29'S	032°06'E	—	05:15:52.6	297	42	30	—			—			07:45:24.6	116	216	63	06:24:19.7	27	130	47	98	0.991	0.992		
Maputo	25°58'S	032°35'E	59	05:18:25.2	298	46	31	—			—			07:45:30.7	114	218	64	06:27:16.2	27	131	46	96	0.965	0.961		
Massingir	23°51'S	032°03'E	—	05:14:53.6	296	42	29	06:22:38.5	149	251	06:23:52.5	264	6	07:41:10.4	117	216	62	06:23:15.4	27	129	45	98	1.005	1.000	0.463	01m14s
Matuba	24°27'S	032°55'E	—	05:16:09.3	296	42	30	06:24:26.6	148	250	06:25:43.5	266	8	07:43:37.5	117	217	64	06:25:05.0	27	129	46	97	1.005	1.000	0.486	01m17s
Nacala	14°34'S	040°41'E	—	05:10:05.2	276	9	34	—			—			07:36:00.0	140	214	67	06:17:57.8	208	295	49	107	0.644	0.560		
Nampula	15°07'S	039°21'E	—	05:09:03.5	279	13	32	—			—			07:35:14.5	137	214	66	06:17:03.0	208	296	48	107	0.683	0.607		
Pafúri	22°27'S	031°21'E	—	05:12:32.8	295	40	28	—			—			07:37:38.1	118	214	61	06:20:18.6	207	307	43	100	0.995	0.996		
Porto Amélia	12°58'S	040°30'E	—	05:08:47.0	274	6	33	—			—			07:31:39.5	142	214	65	06:15:16.7	208	293	48	109	0.607	0.516		
Quelimane	17°53'S	036°51'E	—	05:09:42.1	285	23	31	—			—			07:37:48.1	130	215	65	06:18:36.5	208	300	47	104	0.794	0.744		
Tete	16°13'S	033°35'E	—	05:05:48.1	285	23	27	—			—			07:29:27.8	129	213	60	06:12:41.4	208	300	42	106	0.810	0.764		
Vila Gomes da C...	24°19'S	033°38'E	—	05:16:16.0	296	41	31	06:25:18.2	49	152	06:25:53.3	4	106	07:44:38.7	118	217	64	06:25:35.7	207	309	47	97	1.001	1.000	0.078	00m35s
Vila Pery	19°08'S	033°29'E	—	05:09:02.8	289	30	28	—			—			07:35:03.0	125	214	62	06:17:04.0	207	303	44	103	0.882	0.855		
Vila Trigo de M...	24°36'S	033°00'E	—	05:16:25.3	296	43	31	06:24:53.6	162	265	06:25:56.5	251	354	07:44:01.1	117	217	64	06:25:24.9	27	129	46	97	1.003	1.000	0.286	01m03s
Xai-Xai	25°02'S	033°34'E	—	05:17:20.6	297	43	31	06:26:14.2	167	271	06:27:12.1	246	350	07:45:43.8	116	217	65	06:26:43.1	27	130	47	96	1.002	1.000	0.229	00m58s
NAMIBIA																										
Chefuzwe	17°38'S	024°30'E	—	05:04:28.0	294	36	18	—			—			07:19:54.6	118	209	49	06:07:51.7	206	303	33	105	0.993	0.995		
Linyanti	18°04'S	024°01'E	—	05:04:57.8	295	38	18	—			—			07:19:59.3	117	209	49	06:08:11.5	26	124	33	105	1.006	1.000	0.731	01m07s
Katima Mulilo	17°28'S	024°10'E	—	05:04:11.8	294	36	18	06:07:38.0	132	230	06:08:45.2	281	18	07:19:14.3	118	209	49	06:07:26.0	206	303	32	106	0.995	0.996		
Sikosi	17°59'S	023°19'E	—	05:04:48.5	296	38	17	—			—			07:18:40.0	116	209	48	06:07:40.6	26	124	32	105	0.997	0.998		
Singalamwe	17°41'S	023°23'E	—	05:04:26.1	295	38	18	06:06:46.2	139	237	06:07:49.2	273	11	07:18:37.4	117	209	48	06:07:17.6	26	124	32	106	1.005	1.000	0.608	01m03s
Windhoek	22°34'S	017°06'E	1728	05:11:44.4	306	55	15	—			—			07:17:46.5	104	205	43	06:11:12.1	25	130	28	104	0.800	0.751		
NIGER																										
Maradi	13°29'N	007°06'E	216	—				—			—			06:26:52.0	150	221	10	05:42 Rise	—	—	—	113	0.413	0.299		
Niamey	13°31'N	002°07'E	—	—				—			—			06:28:53.3	145	219	6	05:59 Rise	—	—	—	113	0.374	0.259		
Zinder	13°48'N	008°59'E	—	—				—			—			06:25:32.8	153	223	11	05:39:29.9	203	278	1	113	0.382	0.267		
NIGERIA																										
Enugu	06°27'N	007°27'E	233	—				—			—			06:36:35.7	138	214	15	05:42:15.5	203	285	3	113	0.612	0.520		
Ibadan	07°17'N	003°30'E	—	—				—			—			06:36:07.4	135	213	11	05:46 Rise	—	—	—	112	0.631	0.543		
Iwo	07°38'N	004°11'E	—	—				—			—			06:35:35.7	137	214	11	05:42:27.0	203	285	0	112	0.617	0.526		
Kaduna	10°33'N	007°27'E	—	—				—			—			06:31:04.7	145	218	12	05:40:39.6	203	282	1	113	0.495	0.386		
Kano	12°00'N	008°30'E	—	—				—			—			06:28:36.7	149	220	12	05:40:01.7	203	280	1	113	0.440	0.327		
Lagos	06°27'N	003°24'E	3	—				—			—			06:37:05.5	134	212	12	05:45 Rise	—	—	—	112	0.659	0.576		
Ogbomosho	08°08'N	004°15'E	—	—				—			—			06:34:59.1	138	214	11	05:44 Rise	—	—	—	112	0.601	0.507		
Onitsha	06°09'N	006°47'E	—	—				—			—			06:37:02.7	136	213	15	05:42:31.2	203	286	2	113	0.629	0.540		
Oshogbo	07°47'N	004°34'E	—	—				—			—			06:35:20.8	137	214	12	05:42:17.7	203	285	0	112	0.609	0.516		
Zaria	11°07'N	007°44'E	—	—				—			—			06:30:10.9	146	219	12	05:40:25.7	203	281	1	113	0.475	0.365		
RWANDA																										
Kigali	01°57'S	030°04'E	—	04:56:04.0	267	352	16	—			—			06:55:12.8	148	218	43	05:52:05.0	207	287	29	114	0.510	0.404		
SOMALIA																										
Kismaayo	00°22'S	042°32'E	—	05:13:16.0	249	325	31	—			—			06:53:57.6	168	227	52	06:01:02.8	209	278	41	120	0.245	0.140		
Marka	01°43'N	044°53'E	—	05:22:36.8	240	311	34	—			—			06:43:58.5	178	234	51	06:01:38.6	209	274	42	123	0.145	0.065		
Muqdisho	02°04'N	045°22'E	12	05:24:57.8	238	308	35	—			—			06:41:43.7	180	236	51	06:01:53.6	209	273	43	123	0.127	0.054		

TABLE 15
LOCAL CIRCUMSTANCES FOR AFRICA: SOUTH AFRICA
TOTAL SOLAR ECLIPSE OF 2002 DECEMBER 04

Location Name	Latitude	Longitude	Elev.	First Contact U.T. h m s	P °	V °	Alt °	Second Contact U.T. h m s	P °	V °	Third Contact U.T. h m s	P °	V °	Fourth Contact U.T. h m s	P °	V °	Alt °	Maximum Eclipse U.T. h m s	P °	V °	Alt °	Azm °	Eclip. Mag.	Eclip. Obs.	Umbral Depth	Umbral Durat.
SOUTH AFRICA			m																							
Alexandria	33°39'S	026°24'E	—	05:31:07.4	312	69	30	—			—			07:45:10.3	98	216	57	06:34:35.0	25	141	43	90	0.705	0.633		
Bellville	33°53'S	018°36'E	—	05:32:04.1	318	77	23	—			—			07:32:51.5	90	207	48	06:29:33.0	24	141	35	95	0.591	0.497		
Bloemfontein	29°12'S	026°07'E	—	05:22:30.4	307	60	27	—			—			07:39:09.7	104	213	56	06:26:52.8	26	136	41	95	0.793	0.743		
Cape Town	33°55'S	018°22'E	—	05:32:11.0	318	77	23	—			—			07:32:32.7	90	206	48	06:29:28.4	24	141	35	95	0.587	0.492		
Carletonville	26°23'S	027°22'E	17	05:17:41.9	303	52	26	—			—			07:37:19.4	109	213	57	06:23:14.6	26	132	41	97	0.873	0.843		
Durban	29°55'S	030°56'E	5	05:24:42.3	304	57	32	—			—			07:48:26.0	107	219	63	06:32:18.1	26	136	46	92	0.851	0.817		
East London	33°00'S	027°55'E	—	05:29:58.8	310	66	31	—			—			07:47:02.6	100	217	59	06:34:46.0	25	140	44	90	0.740	0.677		
Edendale	29°39'S	030°18'E	—	05:24:02.9	305	57	31	—			—			07:46:55.9	107	218	62	06:31:14.6	26	136	45	92	0.847	0.811		
Evaton	26°31'S	027°54'E	—	05:18:00.8	303	52	27	—			—			07:38:23.0	109	214	58	06:23:53.8	26	132	41	97	0.878	0.850		
Johannesburg	26°15'S	028°00'E	—	05:17:34.9	302	51	27	—			—			07:38:09.6	110	214	58	06:23:32.6	26	132	41	97	0.885	0.860		
Kimberley	28°43'S	024°46'E	1197	05:21:31.5	308	60	25	—			—			07:36:19.8	103	212	55	06:25:02.9	26	135	39	96	0.783	0.730		
Klerksdorp	26°58'S	026°39'E	—	05:18:35.5	304	54	26	—			—			07:36:59.7	107	213	56	06:23:37.6	26	133	40	97	0.849	0.813		
Kwa-Thema	26°18'S	028°23'E	—	05:17:44.3	302	51	27	—			—			07:38:15.7	110	214	58	06:23:56.6	26	132	42	97	0.890	0.866		
Londolozi	24°42'S	031°31'E	—	05:16:00.0	298	44	29	—			—			07:41:38.2	115	216	62	06:24:07.5	27	130	44	98	0.976	0.975		
Louis Trichardt	23°03'S	029°55'E	—	05:12:57.3	297	42	27	—			—			07:39:19.6	116	214	59	06:19:59.4	27	128	42	100	0.988	0.988		
Mamelodi	25°45'S	028°18'E	—	05:16:48.4	302	50	27	—			—			07:37:54.8	110	214	58	06:22:58.6	26	131	41	98	0.901	0.880		
Mdantsana	32°56'S	027°42'E	—	05:29:49.1	310	66	30	—			—			07:46:34.9	100	217	59	06:34:27.8	25	140	44	90	0.738	0.675		
Messina	22°23'S	030°00'E	—	05:11:59.7	296	41	26	06:18:26.1	152	253	06:19:34.0	262	2	07:35:32.4	116	213	59	06:19:00.0	27	127	41	100	1.004	1.000	0.424	01m08s
Mopane	22°37'S	029°52'E	—	05:12:17.8	297	42	26	—			—			07:35:32.4	116	213	59	06:19:15.3	27	128	41	100	0.997	0.998		
Natalspruit	26°19'S	028°09'E	—	05:17:43.3	302	51	27	—			—			07:38:09.3	110	214	58	06:23:46.6	26	132	41	97	0.886	0.861		
Nelspruit	24°28'S	030°59'E	—	05:15:27.2	298	44	29	—			—			07:40:21.8	115	215	61	06:23:14.5	27	130	44	98	0.973	0.971		
Paarl	33°45'S	018°56'E	—	05:31:44.0	318	76	24	—			—			07:33:13.0	91	207	49	06:29:31.4	24	141	36	95	0.598	0.505		
Parow	33°53'S	018°37'E	—	05:32:08.3	318	77	24	—			—			07:32:53.0	90	207	48	06:29:30.0	24	141	35	95	0.591	0.497		
Pietermaritzburg	29°37'S	030°16'E	—	05:23:58.7	305	57	31	—			—			07:46:49.6	107	218	62	06:31:09.3	26	136	45	93	0.847	0.812		
Pietersburg	23°55'S	029°26'E	—	05:14:08.3	299	45	27	—			—			07:36:56.4	114	214	59	06:20:57.9	27	129	42	99	0.960	0.955		
Pinetown	29°52'S	030°46'E	—	05:24:34.1	304	57	31	—			—			07:48:03.9	107	219	62	06:32:03.3	26	136	46	92	0.850	0.815		
Port Elizabeth	33°58'S	025°40'E	58	05:31:43.3	313	70	29	—			—			07:44:16.7	97	215	57	06:34:31.5	25	141	42	90	0.688	0.613		
Pretoria	25°45'S	028°10'E	1369	05:16:46.0	302	50	27	—			—			07:37:40.9	110	213	58	06:22:51.0	26	131	41	98	0.899	0.877		
Randfontein	26°11'S	027°42'E	—	05:17:25.0	303	52	26	—			—			07:37:34.4	109	213	58	06:23:11.3	26	132	41	98	0.882	0.856		
Sabi Sabi	24°55'S	031°31'E	—	05:16:20.5	298	45	29	—			—			07:41:59.2	115	216	62	06:24:28.9	27	130	44	97	0.971	0.969		
Soweto	26°14'S	027°54'E	—	05:17:32.1	302	52	27	—			—			07:37:58.3	109	213	58	06:23:26.0	26	132	41	97	0.884	0.858		
Tembisa	25°58'S	028°14'E	—	05:17:09.1	302	52	27	—			—			07:38:07.5	110	214	58	06:23:16.4	26	132	41	98	0.895	0.872		
Thohoyandou	23°00'S	030°29'E	—	05:13:03.3	297	42	27	—			—			07:37:09.2	116	214	60	06:20:24.5	27	129	42	100	0.998	0.999		
Tzaneen	23°50'S	030°10'E	—	05:14:13.0	298	44	27	—			—			07:37:59.8	115	214	60	06:21:28.2	27	129	42	99	0.974	0.972		
Uitenhage	33°40'S	025°28'E	—	05:31:06.1	313	70	29	—			—			07:43:35.6	97	214	56	06:33:51.4	25	141	42	91	0.691	0.617		
Vanderbijlpark	26°42'S	027°54'E	—	05:18:19.4	303	53	27	—			—			07:38:38.9	109	214	58	06:24:07.7	26	133	41	97	0.874	0.845		
Vereeniging	26°38'S	027°57'E	—	05:18:13.2	303	53	27	—			—			07:38:38.1	109	214	58	06:24:07.1	26	132	41	97	0.876	0.848		
Welkom	27°59'S	026°45'E	—	05:20:22.8	305	56	27	—			—			07:41:59.2	115	215	57	06:25:22.3	26	134	41	96	0.828	0.787		
SOUTH AFRICA — KRUGER NATIONAL PARK																										
Bateleur Camp	23°14'S	031°12'E	—	05:13:38.6	296	42	28	06:20:50.3	150	251	06:22:02.1	264	5	07:38:42.4	117	215	61	06:21:26.1	27	128	43	99	1.005	1.000	0.452	01m12s
Babalala Picnic	22°54'S	031°15'E	—	05:13:15.8	296	41	28	06:20:16.0	100	201	06:21:38.2	313	54	07:38:14.4	117	215	61	06:20:57.0	207	308	43	99	1.009	1.000	0.715	01m22s
Kanniedood Dam	23°08'S	031°27'E	—	05:13:34.5	296	41	28	06:20:47.0	110	212	06:22:12.7	303	44	07:38:56.6	117	215	61	06:21:29.7	207	308	43	99	1.009	1.000	0.889	01m26s
Letaba Camp	23°51'S	031°35'E	—	05:14:42.8	297	42	28	—			—			07:40:22.4	116	215	61	06:21:29.4	27	129	44	99	0.997	0.998		
Lower Sabie Camp	25°07'S	031°55'E	—	05:16:48.6	298	45	30	—			—			07:42:59.9	115	215	63	06:25:12.2	27	130	45	97	0.973	0.971		
Mooiplaas Picnic	23°33'S	031°26'E	—	05:14:12.8	296	42	28	06:21:51.4	180	282	06:22:30.5	234	335	07:39:38.3	117	215	61	06:22:10.9	27	128	44	99	1.001	1.000	0.108	00m39s
Mopani Camp	23°31'S	031°24'E	—	05:14:08.8	296	42	28	06:21:44.8	179	280	06:22:25.6	235	337	07:39:30.3	117	215	61	06:22:05.1	27	128	44	99	1.001	1.000	0.119	00m41s
Olifants Camp	24°00'S	031°44'E	—	05:15:00.3	297	43	29	—			—			07:40:52.7	116	215	62	06:23:11.8	27	129	44	98	0.996	0.997		
Punda Maria Camp	22°42'S	031°01'E	—	05:12:46.7	296	41	28	06:19:44.5	94	195	06:21:02.9	320	60	07:37:30.1	117	214	60	06:20:23.6	207	308	43	100	1.006	1.000	0.609	01m18s
Punda Maria Gate	22°44'S	031°01'E	—	05:12:50.2	296	41	28	06:19:47.4	98	199	06:21:08.2	315	56	07:37:34.8	117	214	60	06:20:27.7	207	308	43	100	1.007	1.000	0.682	01m21s
Satara Camp	24°24'S	031°47'E	—	05:15:37.0	297	43	29	—			—			07:41:35.1	116	215	62	06:23:52.3	27	129	45	98	0.988	0.989		
Shingwedzi Camp	23°07'S	031°26'E	—	05:13:32.5	296	41	28	06:20:44.4	110	211	06:22:10.0	304	45	07:38:53.1	117	215	61	06:21:27.1	207	308	43	99	1.009	1.000	0.882	01m26s
Sirheni Camp	22°57'S	031°13'E	—	05:13:15.7	296	41	28	06:20:17.1	109	210	06:21:41.9	305	46	07:38:15.7	117	215	61	06:20:59.4	207	308	43	100	1.009	1.000	0.858	01m25s
Skukuza Camp	22°59'S	031°36'E	—	05:16:28.8	298	45	29	—			—			07:42:13.8	115	216	62	06:24:40.2	27	130	45	97	0.971	0.969		
Tshange Lookout	23°14'S	031°14'E	—	05:13:38.8	296	41	28	06:20:49.7	146	247	06:22:05.0	268	9	07:38:45.0	117	215	61	06:21:27.3	27	128	43	99	1.005	1.000	0.518	01m15s

TABLE 16
LOCAL CIRCUMSTANCES FOR AFRICA: SUDAN – ZIMBABWE
TOTAL SOLAR ECLIPSE OF 2002 DECEMBER 04

Location Name	Latitude	Longitude	Elev. (m)	First Contact U.T. (h m s)	P °	V °	Alt °	Second Contact U.T. (h m s)	P °	V °	Third Contact U.T. (h m s)	P °	V °	Fourth Contact U.T. (h m s)	P °	V °	Alt °	Maximum Eclipse U.T. (h m s)	P °	V °	Alt °	Azm °	Eclip. Mag.	Eclip. Obs.	Umbral Depth	Umbral Durat.
SUDAN																										
Al-Fashir	13°38'N	025°21'E	--	05:06:30.4	238	310	8	--	--	--	--	--	--	06:15:38.7	174	237	23	05:40:00.1	206	274	15	118	0.162	0.076		
Al-Ubayyid	13°11'N	030°13'E	--	05:14:05.1	232	301	14	--	--	--	--	--	--	06:10:58.7	182	244	26	05:41:49.4	207	273	20	120	0.098	0.036		
Juba	04°51'N	031°37'E	--	05:00:40.7	252	330	16	--	--	--	--	--	--	06:38:14.6	162	227	37	05:47:06.2	207	280	26	118	0.302	0.191		
Kusti	13°10'N	032°40'E	--	05:20:48.1	226	294	17	--	--	--	--	--	--	06:06:00.2	188	249	27	05:42:59.7	207	276	22	121	0.058	0.017		
Nyala	12°03'N	024°53'E	--	05:02:43.9	243	317	7	--	--	--	--	--	--	06:20:55.9	169	233	24	05:40:25.3	206	276	16	117	0.213	0.115		
Wad Madani	14°25'N	033°28'E	--	05:33:14.9	215	279	20	--	--	--	--	--	--	05:52:39.6	199	261	24	05:42:58.3	207	270	22	122	0.010	0.001		
Waw	07°42'N	028°00'E	--	04:59:43.4	250	327	11	--	--	--	--	--	--	06:31:20.3	163	229	31	05:43:30.2	207	279	21	117	0.285	0.175		
SWAZILAND																										
Mbabane	26°18'S	031°06'E	--	05:18:26.5	300	48	30	--	--	--	--	--	--	07:43:26.7	112	217	62	06:26:22.5	26	132	45	96	0.934	0.921		
TANZANIA																										
Arusha	03°22'S	036°41'E	--	05:02:03.8	263	346	24	--	--	--	--	--	--	07:02:07.8	154	220	51	05:58:26.5	208	285	37	115	0.431	0.318		
Dar es Salaam	06°48'S	039°17'E	14	05:05:05.4	266	351	29	--	--	--	--	--	--	07:13:34.1	151	217	57	06:05:10.6	208	287	42	114	0.472	0.363		
Dodoma	06°11'S	035°45'E	--	05:01:04.8	268	355	24	--	--	--	--	--	--	07:09:02.0	148	215	53	06:00:55.0	208	288	38	113	0.521	0.416		
Iringa	07°46'S	035°42'E	--	05:01:01.5	271	359	24	--	--	--	--	--	--	07:12:59.2	145	216	55	06:02:52.0	208	290	39	112	0.562	0.463		
Mbeya	08°54'S	033°27'E	--	05:00:03.8	275	5	23	--	--	--	--	--	--	07:13:33.0	140	215	53	06:02:21.7	208	292	37	111	0.631	0.544		
Moshi	03°21'S	037°20'E	--	05:02:51.3	262	345	25	--	--	--	--	--	--	07:02:25.6	155	220	52	05:59:00.8	208	285	38	116	0.419	0.306		
Mwanza	02°31'S	032°54'E	--	04:58:16.4	265	350	19	--	--	--	--	--	--	06:57:56.1	146	219	46	05:54:29.5	208	286	32	115	0.476	0.367		
Tabora	05°01'S	032°48'E	--	04:58:17.3	269	3	21	--	--	--	--	--	--	07:03:57.7	140	217	49	05:57:09.0	208	288	34	113	0.543	0.441		
Tanga	05°04'S	039°06'E	--	05:04:51.8	263	347	28	--	--	--	--	--	--	07:08:26.1	154	219	55	06:02:47.2	208	285	41	115	0.431	0.319		
Zanzibar	06°10'S	039°11'E	--	05:04:55.8	265	350	28	--	--	--	--	--	--	07:11:40.9	152	218	57	06:04:15.3	208	286	42	114	0.458	0.347		
TOGO																										
Lomé	06°08'N	001°13'E	22	--	--	--	--	--	--	--	--	--	--	06:37:45.8	132	211	10	05:52 Rise	--	--	0	112	0.658	0.574		
Sokodé	08°59'N	001°08'E	--	--	--	--	--	--	--	--	--	--	--	06:34:42.0	136	213	8	05:58 Rise	--	--	0	112	0.523	0.418		
TRANSKEI																										
Umtata	31°35'S	028°47'E	--	05:27:20.6	308	62	31	--	--	--	--	--	--	07:46:48.7	103	218	60	06:33:07.5	26	138	44	91	0.783	0.730		
ZAMBIA																										
Chililabombwe	12°18'S	027°43'E	--	04:59:30.0	285	21	18	--	--	--	--	--	--	07:14:33.7	129	212	50	06:02:34.8	207	297	33	109	0.813	0.768		
Chingola	12°32'S	027°52'E	--	04:59:45.0	285	21	19	--	--	--	--	--	--	07:15:09.3	129	212	50	06:02:59.0	207	298	33	109	0.817	0.772		
Kabwe	14°27'S	028°27'E	--	05:00:20.7	287	25	20	--	--	--	--	--	--	07:19:20.7	127	212	50	06:05:57.5	207	299	35	107	0.854	0.819		
Kalulushi	12°50'S	028°03'E	--	05:00:04.5	285	22	19	--	--	--	--	--	--	07:15:54.7	129	212	50	06:03:30.0	207	298	34	108	0.821	0.778		
Kitwe	12°49'S	028°13'E	--	05:02:51.3	285	22	19	--	--	--	--	--	--	07:16:04.0	129	212	51	06:03:35.2	207	298	34	108	0.818	0.774		
Livingstone	17°50'S	025°53'E	--	05:04:54.1	294	35	20	--	--	--	--	--	--	07:21:58.6	119	210	51	06:09:00.1	207	303	35	105	0.976	0.974		
Luanshya	13°08'S	028°24'E	--	05:00:27.7	285	22	20	--	--	--	--	--	--	07:16:51.7	128	212	51	06:04:08.0	207	298	34	108	0.822	0.780		
Lusaka	15°25'S	028°17'E	1277	05:02:39.3	289	27	21	--	--	--	--	--	--	07:20:52.4	125	211	53	06:07:10.2	207	300	36	107	0.880	0.853		
Mufulira	12°33'S	028°14'E	--	04:59:53.1	285	21	19	--	--	--	--	--	--	07:15:35.4	129	212	50	06:03:14.7	207	297	34	108	0.811	0.765		
Ndola	12°58'S	028°38'E	--	05:00:23.5	285	21	20	--	--	--	--	--	--	07:16:49.0	129	212	51	06:04:04.1	207	298	34	108	0.815	0.770		
ZIMBABWE																										
Antelope Mine	21°02'S	028°27'E	--	05:09:38.5	296	40	24	06:14:56.7	116	216	06:16:16.2	297	37	07:30:46.2	117	212	56	06:15:36.4	207	306	39	102	1.009	1.000	0.989	01m20s
Beitbridge	22°13'S	030°00'E	--	05:11:45.2	296	41	26	06:18:03.4	127	228	06:19:25.0	286	26	07:35:05.4	117	213	59	06:18:44.1	27	127	41	101	1.008	1.000	0.813	01m22s
Bulawayo	20°09'S	028°36'E	--	05:08:27.1	295	38	24	--	--	--	--	--	--	07:29:30.7	119	212	56	06:14:34.0	207	305	39	103	0.987	0.987		
Chirundu	15°59'S	028°54'E	--	05:03:28.6	289	28	22	--	--	--	--	--	--	07:22:41.4	124	125	54	06:08:26.1	207	301	37	106	0.883	0.857		
Chitungwiza	17°45'S	031°16'E	--	05:06:21.0	289	29	25	--	--	--	--	--	--	07:29:08.6	121	213	58	06:12:54.8	207	302	40	104	0.886	0.861		
Colleen Bawn	21°00'S	029°13'E	--	05:08:13.8	289	29	25	--	--	--	--	--	--	07:31:51.6	118	212	58	06:16:10.1	207	304	40	102	0.998	0.998		
Figtree	20°24'S	028°21'E	--	05:08:44.6	295	38	24	--	--	--	--	--	--	07:29:34.7	118	212	56	06:14:33.4	207	305	39	103	0.996	0.996		
Gwanda	20°57'S	029°01'E	--	05:09:01.8	295	39	25	--	--	--	--	--	--	07:31:28.6	118	211	53	06:15:55.7	207	306	36	107	0.998	0.999		
Gwelo	19°27'S	029°49'E	--	05:07:54.2	293	39	25	--	--	--	--	--	--	07:30:06.3	121	212	57	06:14:16.4	207	304	40	103	0.950	0.942		
Harare	17°50'S	031°03'E	1472	05:06:20.3	290	30	25	--	--	--	--	--	--	07:28:58.3	124	213	58	06:12:50.0	207	302	40	104	0.892	0.868		
Kariba	16°31'S	028°47'E	--	05:04:01.8	290	29	22	--	--	--	--	--	--	07:23:29.7	124	212	54	06:09:06.8	207	306	37	106	0.898	0.875		
Kezi	20°58'S	028°32'E	--	05:09:34.2	296	39	24	06:14:56.4	98	198	06:16:12.1	315	54	07:30:47.0	117	212	56	06:15:34.2	207	306	39	102	1.007	1.000	0.687	01m16s
Legion Mine	21°23'S	028°33'E	--	05:10:09.5	296	40	24	06:15:43.7	158	258	06:16:43.8	255	355	07:31:29.5	117	212	56	06:16:13.6	27	127	39	102	1.003	1.000	0.340	00m38s
Marula	20°26'S	028°06'E	--	05:08:43.7	295	39	23	06:14:05.2	56	155	06:14:44.2	357	96	07:29:16.2	118	212	55	06:14:24.9	207	306	38	103	1.001	1.000	0.128	00m00s
Mazunga	21°45'S	029°52'E	--	05:11:02.6	296	40	26	06:17:23.1	73	173	06:18:23.0	340	80	07:34:06.3	118	213	58	06:17:53.0	207	303	41	101	1.003	1.000	0.313	00m00s
Mutare	18°58'S	032°40'E	--	05:08:26.2	290	30	27	--	--	--	--	--	--	07:33:28.9	124	214	60	06:16:01.7	207	303	43	103	0.892	0.868		
Plumtree	20°30'S	027°50'E	--	05:08:45.5	296	39	23	06:13:41.2	102	201	06:14:56.7	311	50	07:28:59.6	117	211	55	06:14:18.8	207	305	38	103	1.007	1.000	0.747	01m16s
Tjolotjo	19°47'S	027°46'E	--	05:07:46.4	295	38	23	--	--	--	--	--	--	07:27:43.2	118	211	55	06:13:10.7	207	305	38	103	0.992	0.993		
Tuli	21°59'S	029°15'E	--	05:11:11.9	296	41	25	06:17:26.0	181	281	06:18:02.0	233	333	07:33:32.3	117	213	58	06:17:44.0	27	127	40	101	1.001	1.000	0.103	00m36s
Victoria Falls	17°50'S	025°51'E	--	05:04:53.7	294	35	20	--	--	--	--	--	--	07:21:56.0	119	210	51	06:08:58.8	207	303	34	105	0.977	0.975		

TABLE 17
LOCAL CIRCUMSTANCES FOR ANTARCTICA, INDONESIA & INDIAN OCEAN
TOTAL SOLAR ECLIPSE OF 2002 DECEMBER 04

Location Name	Latitude	Longitude	Elev.	First Contact U.T. h m s	P °	V °	Alt °	Second Contact U.T. h m s	P °	V °	Third Contact U.T. h m s	P °	V °	Fourth Contact U.T. h m s	P °	V °	Alt °	Maximum Eclipse U.T. h m s	P °	V °	Alt °	Azm °	Eclip. Mag.	Eclip. Obs.	Umbral Depth	Umbral Durat.
ANTARCTICA			m																							
Asuka – JPN	71°32'S	024°08'E	–	07:11:06.4	357	161	34	–			–			07:54:39.0	37	204	37	07:32:47.1	17	183	36	47	0.062	0.018		
Casey – AUS	66°17'S	110°32'E	–	07:35:55.3	324	165	37	–			–			09:10:03.0	58	264	28	08:23:42.2	11	215	33	290	0.331	0.218		
Davis – AUS	68°35'S	077°58'E	–	07:14:50.6	329	154	43	–			–			08:51:05.1	60	255	39	08:03:11.8	14	205	42	333	0.306	0.194		
Dumont d'Urvill...	66°35'S	140°00'E	–	07:53:51.5	332	178	24	–			–			09:04:24.1	47	251	17	08:29:33.3	9	215	20	261	0.208	0.111		
Mawson – AUS	67°40'S	063°43'E	–	07:04:58.3	331	147	44	–			–			08:41:31.6	60	249	43	07:53:13.1	16	198	44	354	0.297	0.187		
Mirny – RUS	66°35'S	093°00'E	–	07:22:58.1	324	157	43	–			–			09:04:23.2	62	264	35	08:14:16.2	13	212	39	311	0.357	0.243		
Molodezhnaja – ...	67°35'S	046°35'E	–	06:58:12.7	338	145	42	–			–			08:14:46.5	56	234	45	07:40:51.5	17	189	44	20	0.232	0.130		
Syowa – JPN	69°00'S	039°35'E	–	07:00:58.5	343	149	39	–			–			08:14:46.5	51	224	42	07:37:37.8	17	186	41	29	0.173	0.084		
Vostok – RUS	78°45'S	107°27'E	–	07:43:16.1	350	179	30	–			–			08:30:49.6	36	226	28	08:07:06.0	13	203	29	304	0.081	0.027		
Zhongshan – CHN	69°22'S	076°23'E	–	07:15:08.4	331	154	43	–			–			08:48:14.4	58	252	39	08:01:52.2	14	203	41	336	0.285	0.176		
COMOROS																										
Moroni	11°41'S	043°16'E	–	05:11:36.9	269	358	35	–			–			07:31:47.4	147	213	67	06:16:56.8	209	290	50	111	0.523	0.420		
INDONESIA																										
Banyuwangi	08°12'S	114°21'E	–	08:48:53.7	211	121	22	–			–			09:48:18.8	157	62	9	09:19:28.2	184	91	15	249	0.105	0.041		
Bojonegoro	07°09'S	111°52'E	–	08:57:04.9	202	113	22	–			–			09:38:00.8	167	74	13	09:17:52.7	184	94	18	249	0.046	0.012		
Cilacap	07°44'S	109°00'E	–	08:57:15.5	199	111	25	–			–			09:32:52.3	170	78	17	09:15:16.3	185	95	21	249	0.032	0.007		
Genteng	08°22'S	114°09'E	–	08:48:16.8	211	121	23	–			–			09:48:33.1	157	61	9	09:19:18.6	184	91	16	249	0.108	0.042		
Jember	08°10'S	113°42'E	–	08:49:28.6	210	120	23	–			–			09:47:01.1	158	63	10	09:19:03.1	184	92	16	249	0.097	0.036		
Jombang	07°33'S	112°14'E	–	08:54:01.1	205	116	23	–			–			09:41:12.3	164	70	12	09:18:06.3	184	93	17	249	0.062	0.018		
Kebumen	07°40'S	109°39'E	–	08:56:41.5	200	112	25	–			–			09:34:35.8	169	77	16	09:15:54.2	185	95	20	249	0.037	0.009		
Klaten	07°42'S	110°35'E	–	08:55:03.8	202	114	24	–			–			09:37:40.1	166	74	15	09:16:44.3	184	94	19	249	0.048	0.013		
Muncar	08°26'S	114°20'E	–	08:47:52.9	212	122	23	–			–			09:49:05.7	156	61	9	09:19:24.9	184	91	16	249	0.112	0.044		
Pare	07°46'S	112°11'E	–	08:52:48.0	206	117	23	–			–			09:42:09.4	163	69	12	09:18:01.9	184	93	18	249	0.068	0.021		
Pati	06°45'S	111°01'E	–	09:01:57.5	197	109	22	–			–			09:32:18.0	171	80	15	09:17:14.2	184	95	18	249	0.024	0.005		
Pemalang	06°54'S	109°22'E	–	09:05:21.5	193	105	23	–			–			09:26:11.8	176	86	18	09:15:44.1	185	95	20	249	0.011	0.001		
Purwokerto	07°25'S	109°14'E	–	08:59:42.0	198	109	24	–			–			09:31:08.2	171	80	17	09:15:32.6	185	95	21	249	0.025	0.005		
Semarang	06°58'S	110°25'E	–	09:01:19.3	197	109	23	–			–			09:31:50.7	171	80	16	09:16:41.6	184	95	19	249	0.024	0.005		
Sidoarjo	07°27'S	112°43'E	–	08:54:04.4	205	116	23	–			–			09:41:52.2	163	70	11	09:18:28.9	184	93	17	249	0.064	0.019		
Surabaya	07°15'S	112°45'E	–	08:55:13.8	204	115	22	–			–			09:40:56.1	164	71	12	09:18:32.1	184	93	17	249	0.059	0.017		
Taman	07°25'S	112°41'E	–	08:54:18.4	205	116	23	–			–			09:41:37.5	163	70	12	09:18:27.7	184	93	17	249	0.063	0.019		
Tasikmalaya	07°20'S	108°12'E	–	09:03:40.6	194	106	24	–			–			09:25:25.9	176	86	19	09:14:31.2	185	96	22	249	0.012	0.002		
MAURITIUS																										
Curepipe	20°19'S	057°31'E	–	05:43:00.5	267	360	58	–			–			08:22:37.5	145	75	84	06:58:43.5	206	291	76	100	0.498	0.392		
Port Louis	20°10'S	057°30'E	55	05:42:54.2	266	359	58	–			–			08:22:08.4	145	77	84	06:58:25.7	206	290	76	101	0.494	0.388		
Rose Hill	20°14'S	057°27'E	–	05:42:48.6	267	359	58	–			–			08:22:14.0	145	76	85	06:58:25.1	206	291	75	101	0.497	0.391		
Vacoas	20°18'S	057°29'E	–	05:42:55.2	267	360	58	–			–			08:22:30.2	145	75	84	06:58:36.9	206	291	76	100	0.498	0.392		
REUNION																										
Saint-Denis	20°52'S	055°28'E	936	05:38:33.5	270	5	55	–			–			08:19:49.9	142	80	87	06:54:44.7	207	295	73	98	0.546	0.446		
SEYCHELLES																										
Victoria	04°38'S	055°27'E	5	05:48:08.9	236	304	52	–			–			07:10:32.8	181	222	68	06:27:47.7	209	266	60	128	0.110	0.043		

TABLE 18
LOCAL CIRCUMSTANCES FOR AUSTRALIA & NEW ZEALAND
TOTAL SOLAR ECLIPSE OF 2002 DECEMBER 04

Location Name	Latitude	Longitude	Elev. m	First Contact U.T. h m s	P °	V °	Alt °	Second Contact U.T. h m s	P °	V °	Third Contact U.T. h m s	P °	V °	Fourth Contact U.T. h m s	P °	V °	Alt °	Maximum Eclipse U.T. h m s	P °	V °	Alt °	Azm °	Eclip. Mag.	Eclip. Obs.	Umbral Depth	Umbral Durat.
AUSTRALIA																										
Adelaide, AS	34°55'S	138°35'E	6	08:09:37.6	282	161	18	—	—	—	—	—	—	—	—	—	—	09:07:31.0	5	240	7	247	0.882	0.853		
Alice Springs	23°42'S	133°53'E	546	08:18:41.0	261	152	17	—	—	—	—	—	—	—	—	—	—	09:16:43.4	184	70	4	247	0.781	0.725		
Andamooka, AS	30°27'S	137°12'E	—	08:12:53.3	274	158	17	—	—	—	—	—	—	—	—	—	—	09:11:28.0	184	64	5	247	0.998	0.999		
Bankstown, SW	33°55'S	151°02'E	—	08:12:06.3	288	164	7	—	—	—	—	—	—	—	—	—	—	08:54 Set	—	—	0	242	0.697	0.618		
Blacktown, SW	33°46'S	150°55'E	—	08:12:11.2	288	164	7	—	—	—	—	—	—	—	—	—	—	08:54 Set	—	—	0	242	0.694	0.621		
Bransby, QL	28°14'S	142°04'E	—	08:15:31.6	274	158	12	—	—	—	—	—	—	—	—	—	—	09:12:01.1	184	64	1	245	0.993	0.992		
Brisbane, QL	27°28'S	153°02'E	—	08:15:32.2	279	161	3	—	—	—	—	—	—	—	—	—	—	08:31 Set	—	—	0	244	0.285	0.175		
Broadmeadows, VC	37°40'S	144°54'E	5	08:09:16.3	290	165	14	—	—	—	—	—	—	—	—	—	—	09:03:42.4	5	236	5	244	0.752	0.689		
Cairns, QL	16°55'S	145°46'E	—	08:24:33.1	258	151	3	—	—	—	—	—	—	—	—	—	—	08:39 Set	—	—	0	246	0.243	0.138		
Campbelltown, SW	34°04'S	150°49'E	—	08:12:01.2	288	164	7	—	—	—	—	—	—	—	—	—	—	08:55 Set	—	—	0	242	0.709	0.637		
Canberra, SW	35°17'S	149°08'E	575	08:11:16.7	289	164	9	—	—	—	—	—	—	—	—	—	—	09:04:22.5	5	236	0	242	0.765	0.705		
Canterbury, SW	33°55'S	151°07'E	—	08:12:06.3	288	164	7	—	—	—	—	—	—	—	—	—	—	08:53 Set	—	—	0	242	0.690	0.613		
Ceduna, AS	32°07'S	133°40'E	—	08:08:11.2	275	158	21	09:10:12.2	95	334	09:10:44.8	274	153	—	—	—	—	09:10:28.5	5	244	3	246	1.005	1.000	0.986	00m33s
Cockburn, SW	32°05'S	141°00'E	—	08:12:23.6	279	160	14	—	—	—	—	—	—	—	—	—	—	09:09:25.6	3	241	6	248	1.000	1.000		
Coondambo, AS	31°04'S	135°32'E	—	08:11:58.9	275	158	18	—	—	—	—	—	—	—	—	—	—	09:11:09.5	185	64	6	248	1.000	1.000		
Denial Bay, AS	32°06'S	133°32'E	—	08:10:15.0	275	158	21	09:10:14.5	73	312	09:10:45.0	296	175	—	—	—	—	09:10:29.8	185	64	9	249	1.003	1.000	0.632	00m30s
Doncaster, VC	37°47'S	145°08'E	—	08:09:14.6	290	165	14	—	—	—	—	—	—	—	—	—	—	09:03:32.4	5	236	4	244	0.747	0.683		
Ediacara, AS	30°24'S	137°54'E	—	08:13:06.1	275	158	16	09:11:09.7	97	336	09:11:37.1	272	151	—	—	—	—	09:11:23.4	4	244	5	247	1.004	1.000	0.951	00m27s
Fairfield, SW	33°52'S	150°57'E	—	08:12:07.9	288	164	7	—	—	—	—	—	—	—	—	—	—	08:54 Set	—	—	0	242	0.698	0.622		
Fort Grey, SW	29°05'S	141°14'E	—	08:14:46.1	275	158	13	09:11:29.0	128	7	09:11:48.6	240	119	—	—	—	—	09:11:38.8	4	243	2	245	1.002	1.000	0.440	00m20s
Geelong, VC	38°08'S	144°21'E	—	08:08:51.7	290	165	14	—	—	—	—	—	—	—	—	—	—	09:03:25.8	5	236	4	245	0.747	0.683		
Glen Waverley	37°53'S	145°10'E	—	08:09:11.3	290	165	14	—	—	—	—	—	—	—	—	—	—	09:03:26.5	5	236	4	244	0.744	0.680		
Glendambo, AS	30°58'S	135°43'E	—	08:12:01.5	274	158	18	—	—	—	—	—	—	—	—	—	—	09:11:15.4	185	64	7	248	0.996	0.997		
Gosford, SW	33°26'S	151°21'E	—	08:12:22.1	287	164	7	—	—	—	—	—	—	—	—	—	—	08:51 Set	—	—	0	243	0.662	0.579		
Hobart, TS	42°53'S	147°19'E	54	08:06:43.9	299	168	14	—	—	—	—	—	—	—	—	—	—	08:58:08.7	6	231	5	244	0.611	0.519		
Keilor, VC	37°43'S	144°50'E	—	08:09:13.6	290	165	14	—	—	—	—	—	—	—	—	—	—	09:03:40.8	7	243	4	245	0.752	0.689		
Leigh Creek, AS	30°28'S	138°25'E	—	08:13:09.7	275	158	16	—	—	—	—	—	—	—	—	—	—	09:11:14.7	4	243	4	247	0.997	0.997		
Logan, QL	27°43'S	153°18'E	—	08:15:20.4	280	161	3	—	—	—	—	—	—	—	—	—	—	08:30 Set	—	—	0	244	0.272	0.163		
Lyndhurst, AS	30°17'S	138°21'E	—	08:13:18.1	275	158	16	09:11:12.2	124	3	09:11:35.7	245	124	—	—	—	—	09:11:24.0	4	243	4	247	1.002	1.000	0.509	00m23s
Melbourne, VC	37°49'S	144°58'E	35	08:09:11.4	290	165	14	—	—	—	—	—	—	—	—	—	—	09:03:33.3	5	236	4	244	0.748	0.684		
Moolawatana, AS	29°55'S	139°43'E	—	08:13:52.4	275	158	14	—	—	—	—	—	—	—	—	—	—	09:11:24.0	4	243	3	246	0.997	0.997		
Mt. Freeling, AS	29°57'S	139°13'E	—	08:13:45.3	275	158	15	09:11:19.4	137	16	09:11:38.5	232	111	—	—	—	—	09:11:29.0	4	243	3	246	1.001	1.000	0.323	00m19s
Newcastle, SW	32°56'S	151°46'E	—	08:12:37.9	287	164	6	—	—	—	—	—	—	—	—	—	—	08:48 Set	—	—	0	243	0.619	0.527		
Nunjikompita, AS	32°16'S	134°19'E	—	08:10:23.8	276	158	20	—	—	—	—	—	—	—	—	—	—	09:10:18.4	5	243	8	249	0.994	0.994		
Parramatta, SW	33°49'S	151°00'E	—	08:12:09.6	288	164	7	—	—	—	—	—	—	—	—	—	—	08:54 Set	—	—	0	242	0.694	0.617		
Penong, AS	31°55'S	133°01'E	—	08:10:12.9	274	158	21	—	—	—	—	—	—	—	—	—	—	09:10:40.5	185	64	9	249	0.993	0.993		
Penrith, SW	33°45'S	150°42'E	—	08:12:11.0	287	164	7	—	—	—	—	—	—	—	—	—	—	08:55 Set	—	—	0	242	0.708	0.635		
Perth, AW	31°57'S	115°51'E	20	07:58:10.6	266	152	38	—	—	—	09:06:38.4	—	—	10:07:36.1	106	346	11	09:06:38.4	186	69	24	258	0.824	0.780		
Pimba, AS	31°15'S	136°47'E	—	08:12:05.3	275	158	18	—	—	—	—	—	—	—	—	—	—	09:10:53.5	5	243	6	247	0.994	0.994		
Port Augusta, AS	32°30'S	137°46'E	—	08:11:18.7	278	159	17	—	—	—	—	—	—	—	—	—	—	09:09:43.2	5	242	6	247	0.952	0.941		
Purple Downs, AS	30°47'S	136°53'E	—	08:12:31.1	275	158	17	09:11:00.9	85	324	09:11:29.2	284	163	—	—	—	—	09:11:15.1	185	64	6	247	0.997	1.000	0.836	00m28s
Randwick, SW	33°55'S	151°15'E	—	08:12:06.3	288	164	7	—	—	—	—	—	—	—	—	—	—	08:53 Set	—	—	0	242	0.682	0.604		
Roxby Downs, AS	30°33'S	136°53'E	—	08:10:43.2	274	158	17	09:11:26.2	158	64	—	—	—	—	—	—	—	09:11:26.2	185	64	5	247	0.997	0.998		
Salisbury, AS	34°46'S	138°38'E	—	08:09:45.4	282	161	18	—	—	—	—	—	—	—	—	—	—	09:10:38.5	5	240	6	247	0.885	0.857		
Smoky Bay, AS	32°22'S	133°56'E	—	08:10:09.3	276	158	21	—	—	—	—	—	—	—	—	—	—	09:10:14.8	5	243	9	249	0.995	0.996		
Southport, QL	27°58'S	153°25'E	—	08:15:25.8	280	161	3	—	—	—	—	—	—	—	—	—	—	08:30 Set	—	—	0	244	0.276	0.167		
Sydney, SW	33°52'S	151°13'E	19	08:12:08.0	288	164	7	—	—	—	—	—	—	—	—	—	—	08:54 Set	—	—	0	242	0.694	0.617		
Thevenard, AS	32°09'S	133°38'E	—	08:10:14.6	275	158	21	09:10:10.7	102	341	09:10:43.1	268	147	—	—	—	—	09:10:26.9	5	244	4	246	1.004	1.000	0.877	00m32s
Ticklara, QL	28°37'S	140°13'E	—	08:15:14.3	275	158	12	09:11:31.4	105	344	09:11:53.5	263	142	—	—	—	—	09:11:42.5	4	243	1	245	1.003	1.000	0.811	00m22s
Tilcha, AS	29°36'S	140°54'E	—	08:14:18.9	275	158	13	—	—	—	—	—	—	—	—	—	—	09:11:21.4	4	243	2	245	0.992	0.992		
Townsville, QL	19°16'S	146°48'E	—	08:22:17.9	262	153	3	—	—	—	—	—	—	—	—	—	—	08:39 Set	—	—	0	246	0.293	0.182		
Wandana, AS	32°04'S	133°49'E	—	08:10:23.4	275	158	21	09:10:30.6	99	338	09:10:46.7	271	150	—	—	—	—	09:10:30.6	5	244	4	246	1.004	1.000	0.931	00m32s
Wanneroo, AW	31°45'S	115°48'E	—	07:58:22.3	266	152	38	—	—	—	—	—	—	10:07:42.4	106	346	11	09:06:47.7	186	70	24	258	1.001	1.000		
Wirramima, AS	31°12'S	136°15'E	—	08:11:58.7	275	158	18	09:10:50.8	145	24	09:11:09.5	224	103	—	—	—	—	09:11:00.2	5	243	6	248	1.001	1.000	0.229	00m19s
Wollongong, SW	34°25'S	150°54'E	—	08:11:49.7	288	164	8	—	—	—	—	—	—	—	—	—	—	08:55 Set	—	—	0	242	0.712	0.640		
Woomera, AS	31°11'S	136°49'E	—	08:12:09.3	275	158	18	—	—	—	—	—	—	—	—	—	—	09:10:56.4	5	243	6	247	0.995	0.996		
NEW ZEALAND																										
Dunedin	45°52'S	170°30'E	1	08:07:16.7	317	177	1	—	—	—	—	—	—	—	—	—	—	08:13 Set	—	—	0	236	0.086	0.030		
Invercargill	46°24'S	168°21'E	—	08:07:17.6	317	176	2	—	—	—	—	—	—	—	—	—	—	08:24 Set	—	—	0	236	0.216	0.116		
Palmerston	45°29'S	170°43'E	—	08:07:18.4	317	176	0	—	—	—	—	—	—	—	—	—	—	08:11 Set	—	—	0	236	0.050	0.013		

TABLE 19

SOLAR ECLIPSES OF SAROS SERIES 142

First Eclipse: 1624 Apr 17   Duration of Series: 1280.1 yrs.
Last Eclipse: 2904 Jun 05   Number of Eclipses: 72

Saros Summary: Partial: 27   Annular: 1   Total: 43   Hybrid: 1

Date	Eclipse Type	Gamma	Mag./ Width	Center Durat.	Date	Eclipse Type	Gamma	Mag./ Width	Center Durat.
1624 Apr 17	Pb	-1.520	0.058	—	2345 Jun 30	T	0.326	272	06m07s
1642 Apr 29	P	-1.458	0.166	—	2363 Jul 12	T	0.401	279	05m51s
1660 May 09	P	-1.389	0.287	—	2381 Jul 22	T	0.474	285	05m33s
1678 May 20	P	-1.317	0.416	—	2399 Aug 02	T	0.548	291	05m14s
1696 May 30	P	-1.240	0.554	—	2417 Aug 13	T	0.618	297	04m55s
1714 Jun 12	P	-1.161	0.698	—	2435 Aug 24	T	0.687	303	04m35s
1732 Jun 22	P	-1.080	0.846	—	2453 Sep 03	T	0.751	311	04m15s
1750 Jul 03	A-	-0.998	-	-	2471 Sep 15	T	0.810	323	03m54s
1768 Jul 14	H	-0.917	48	00m29s	2489 Sep 25	T	0.865	340	03m32s
1786 Jul 25	T	-0.838	66	00m59s	2507 Oct 07	T	0.913	373	03m07s
1804 Aug 05	T	-0.762	75	01m20s	2525 Oct 18	T	0.955	447	02m39s
1822 Aug 16	T	-0.690	80	01m35s	2543 Oct 29	Tn	0.991	-	02m03s
1840 Aug 27	T	-0.622	83	01m45s	2561 Nov 08	P	1.021	0.966	—
1858 Sep 07	T	-0.561	85	01m50s	2579 Nov 20	P	1.046	0.919	—
1876 Sep 17	T	-0.505	86	01m53s	2597 Nov 30	P	1.065	0.882	—
1894 Sep 29	T	-0.457	86	01m55s	2615 Dec 12	P	1.080	0.853	—
1912 Oct 10	T	-0.415	85	01m55s	2633 Dec 23	P	1.090	0.832	—
1930 Oct 21	T	-0.380	84	01m55s	2652 Jan 03	P	1.099	0.816	—
1948 Nov 01	T	-0.352	84	01m56s	2670 Jan 13	P	1.105	0.803	—
1966 Nov 12	T	-0.330	84	01m57s	2688 Jan 25	P	1.113	0.787	—
1984 Nov 22	T	-0.313	85	02m00s	2706 Feb 05	P	1.121	0.772	—
2002 Dec 04	T	-0.302	87	02m04s	2724 Feb 16	P	1.132	0.752	—
2020 Dec 14	T	-0.294	90	02m10s	2742 Feb 27	P	1.146	0.726	—
2038 Dec 26	T	-0.288	95	02m18s	2760 Mar 09	P	1.166	0.689	—
2057 Jan 05	T	-0.284	102	02m29s	2778 Mar 21	P	1.190	0.645	—
2075 Jan 16	T	-0.280	110	02m42s	2796 Mar 31	P	1.221	0.589	—
2093 Jan 27	T	-0.274	119	02m58s	2814 Apr 11	P	1.258	0.522	—
2111 Feb 08	T	-0.265	130	03m17s	2832 Apr 22	P	1.302	0.441	—
2129 Feb 18	T	-0.253	142	03m38s	2850 May 03	P	1.353	0.349	—
2147 Mar 02	T	-0.236	155	04m02s	2868 May 13	P	1.410	0.245	—
2165 Mar 12	T	-0.213	168	04m27s	2886 May 24	P	1.473	0.130	—
2183 Mar 23	T	-0.185	181	04m54s	2904 Jun 05	Pe	1.542	0.006	—
2201 Apr 04	T	-0.150	194	05m20s					
2219 Apr 15	T	-0.109	207	05m45s					
2237 Apr 25	T	-0.061	219	06m05s					
2255 May 07	T	-0.008	230	06m22s					
2273 May 17	Tm	0.051	240	06m31s					
2291 May 28	T	0.115	249	06m34s					
2309 Jun 09	T	0.183	257	06m30s					
2327 Jun 20	T	0.254	265	06m21s					

Eclipse Type: P - Partial   Pb - Partial Eclipse (Saros Series Begins)
                A- - Non-central Annular   Pe - Partial Eclipse (Saros Series Ends)
                T - Total   Tm - Middle eclipse of Saros series.
                H - Hybrid (Annular/Total)   Tn - Total Eclipse (no northern limit).

Note: Mag./Width column gives either the eclipse magnitude (for partial eclipses)
       or the umbral path width in kilometers (for total and annular eclipses).

Table 20

African and Australian Weather Statistics for the 2002 Total Solar Eclipse

Location	Lat	Long	Sunshine (hours)	Percent of possible sunshine	Mean Cloud Cover	Pcpn Amount (mm)	Mean Days with rain	Tmax (°C)	Tmin (°C)	Days with TRW	Clear (%)	Scattered (%)	Broken (%)	Overcast (%)	Clear and Scattered (%)	Days with Sky <3/10 & good visibiltiy at eclipse time
Angola																
Lubango	-14.77	13.57	6.5	50	0.63	153.0	14.0	25.0	13.0	8						3.9
Lobito	-12.37	13.53				61.0	5.3	31.9	25.0	1						7.8
Cuima	-13.25	15.65			0.63											
Chitembo	-13.52	16.77			0.75											
Huambo	-12.80	15.75	4.5	35	0.75	233.0	17.0	25.6	14.4	11						1.1
Namibia																
Rundu	-17.95	19.72				84.7	5.0	34.4	24.4	5	5.9	34.4	44.6	15.2	40.3	
Gobabeb	-23.55	15.03	10.9	80												
Grootfontein	-19.57	18.12	10.7	76	0.30	81.3	7.0	34.4	22.5	8	7.2	41.8	43.6	7.4	49.0	
Tsumeb	-19.23	17.72	9.5	72	0.40	97.0	11.0	32.0	18.0							
Botswana																
Francistown	-21.17	27.52				89.7	6.0	33.8	22.5	6	12.9	34.8	28.7	23.5	47.7	
Maun	-19.98	23.42	6.9	52	0.63	79.9	6.1	32.8	18.9							4.9
Kasane	-17.82	25.15				146.3										
Shakawe	-18.38	21.85				98.5										
Zambia																
Livingstone	-17.82	29.82	6.7	37	0.75	169.1	12.0	30.4	18.9	21						0.8
Mongu	-15.25	23.17	6.3	48	0.78	192.8	17.0	29.3	18.6	23						0.7
Lusaka	-15.42	28.32	5.6	43	0.75	186.0	16.0	27.0	17.0	21	1	15.1	61	22.7	16.1	0.6
Zimbabwe																
Bulawayo	-20.15	28.62	6.9	52	0.63	128.0	10.0	27.0	16.0	12	1.9	30	58.6	8.5	31.9	3.6
Beitbridge	-22.22	30.00	8.0	59	0.58	58.7	5.0	33.0	21.0	7						
Hwange	-18.37	26.48			0.69	129.8										
Victoria Falls Airport	-18.10	25.85	6.9	52		168.5										
Plumtree	-20.48	27.82				119.8										
Gweru	-19.45	29.85				157.5	16.0	29.4	18.1	10	1.2	41.7	44.8	12.4	42.9	
South Africa																
Johannesburg	-26.13	28.23	8.5	62		105.0	14.0	25.2	13.9	12	7.5	40.3	41.5	10.4	47.8	10.5
Durban	-29.97	30.95	6.1	43	0.60	102.0	17.0	26.9	20.0	4	5.1	28.6	41.3	24.8	33.7	7.0
Skukuza	-24.98	31.60	6.7	49	0.64	92.0		31.9	19.7	3.3						
Ellisras	-24.68	27.68	7.6	56	0.56	80.0		32.2	19.5							
Pietersburg	-24.87	29.45	7.9	58	0.64	81.0	8.0	27.4	16.4	7.8	5.1	27.4	46.2	21.3	32.5	10.3
Tzaneen	-24.83	30.15	6.4	47	0.69	147.0		28.7	18.4	2.2						
Phalaborwa	-24.93	31.15			0.68	99.0		31.1	19.9	1.2						
Marnitz	-24.15	28.22	8.2	60	0.51	56.0		30.8	18.7	7.4						
Mara	-24.15	29.57	7.4	54	0.66	78.0		29.6	17.3	5.3						
Levubu	-24.08	30.28	6.7	49	0.70	136.0		28.7	18.5	3.5						
Messina	-24.27	29.90	8.5	62	0.58	57.0		32.9	20.4	3.1						
Mopani, KNP	-22.60	29.85			0.61	84.7	7.4				19.6	20.1	26.8	33.5	39.7	
Shingwedzi, KNP	-23.12	31.43			0.51	88.3	6.5				18.5	34.5	28.2	18.8	53.0	
Punda Maria, KNP	-22.68	31.02			0.66	92.5	7.6				28.0	12.2	17.3	42.4	40.2	
Pretoria	-25.73	28.18	9.0	66	0.50	111.8	8.0	30.6	20.6	8	9.5	37.9	35.8	16.8	47.4	10.5
Mozambique																
Beira	-19.83	34.85	7.6	58	0.75	234.0	10.0	31.0	23.0	5	1.4	25.2	57.3	16	26.6	7.1
Pafuri	-22.43	31.33	7.0	52	0.63	380.0	7.0	35.0	21.0							
Panda	-24.05	34.72			0.63	108.0	6.0	33.0	19.0							
Inhambane	-23.87	35.38				107.7	8.5	30.6	22.2	6.8						6.6
Maputo	-25.97	32.60	7.1	52	0.88	103.0	8.0	29.0	21.0	2	4.5	31.1	44.6	19.8	35.6	5.2
Others																
Alfred Faure/Crozet	-46.43	51.87									0.2	15.9	51.1	27	16.1	
Martin de Vivies	-37.80	77.80									0	24.1	52.5	22.9	24.1	
Port aux Francais	-49.35	70.25									0	19	66.2	14.6	19.0	
Australia																
Biloela	-24.38	150.52	8.9	65		98.6	8.4	32.9	18.4							
Broken Hill	-31.98	141.47				21.9	3.4	31.3	17.0	2	34.3	29.1	30.1	6.5	63.4	
Bundaberg Aero	-24.91	152.32				127.0	10.1	29.1	20.3	3	11.8	37.9	29.9	20.5	49.7	
Bundaberg PO	-24.87	152.35				131.0	9.6	29.8	20.6	2	19.8	33.3	26.2	20.7	53.1	
Ceduna AMO	-32.13	133.71	9.5	67	0.45	20.5	5.2	27.2	13.9	2	25.1	33.1	29.7	12.1	58.2	13.3
Charleville Aero	-26.41	146.26				53.0	7.0	34.7	20.2	5	16.4	39.6	30.1	13.8	56.0	12.3
Kyancutta	-33.13	135.56	9.1	64		20.5	5.0	30.8	12.8							
Leigh Creek Aero	-30.47	138.41			0.36	21.5	3.0	33.3	18.7	1	37.7	28.9	23.6	9.8	66.6	18.2
Minnipa	-32.84	135.15	9.2	64		19.7	3.9	29.3	13.9							
Marree	-28.65	138.06			0.33	16.5	2.4	36.0	19.4	1						
Oodnadatta	-27.58	135.45	11.0	79	0.30	14.7	3.0	36.3	21.1	1.5	31.4	39.2	26.2	3.2	70.6	16.8
Theodore	-24.84	149.80	7.9	58		104.7	7.5	33.1	19.8							
Todd River	-34.49	135.85	8.9	62		24.7	6.3	25	12.4							
Woomera	-31.15	136.82	10.5	74	0.39	13.7	3.3	32.2	17.5	2	32.7	29.1	29.8	8.4	61.8	16

Key to Table 20 appears on following page...

Key to Table 20

African and Australian Weather Statistics
for the 2002 Total Solar Eclipse

Latitude

Longitude

Sunshine (hours)
Mean daily sunshine for December.

Percent of Possible Sunshine
The ratio of mean daily sunshine hours to the length of the day from sunrise to sunset in mid December. This statistic is one of the most suitable for assessing the chances of seeing the eclipse.

Mean Cloud Cover
Mean cloudiness as measured from ground-based stations. For Australian stations this is the value at 6 PM local time. For other sites, it is the mean cloudiness through the day.

Precipitation Amount (mm)
Average precipitation in mm for December.

Mean days with rain
Average number of days in December with 0.2 mm of rain or greater.

Tmax
Mean daily high temperature for December

Tmin
Mean daily minimum temperature for December

Days with TRW
Average number of days in December with thunderstorms.

Clear, scattered, broken, overcast
Percent frequency of days in December with each category of cloud cover at or close to eclipse time (within 2 hours). Scattered cloud refers to 1-4 tenths and broken to 5-9 tenths.

Clear and scattered cloud
The sum of the percent number of days with clear and scattered cloud. Observers may observe thin clouds as either clear or scattered. Combining the categories reduces individual differences and affords a better comparison between stations.

Days with Sky <3/10 and good visibility at eclipse time
The number of days in December at or close to eclipse time with less than 3-10ths cloud cover and visibilities better than 3 miles. This statistic is also a good measure of the relative cloudiness between stations.

TABLE 21
35 MM FIELD OF VIEW AND SIZE OF SUN'S IMAGE FOR VARIOUS PHOTOGRAPHIC FOCAL LENGTHS

Focal Length	Field of View	Size of Sun
28 mm	49° x 74°	0.2 mm
35 mm	39° x 59°	0.3 mm
50 mm	27° x 40°	0.5 mm
105 mm	13° x 19°	1.0 mm
200 mm	7° x 10°	1.8 mm
400 mm	3.4° x 5.1°	3.7 mm
500 mm	2.7° x 4.1°	4.6 mm
1000 mm	1.4° x 2.1°	9.2 mm
1500 mm	0.9° x 1.4°	13.8 mm
2000 mm	0.7° x 1.0°	18.4 mm
2500 mm	0.6° x 0.8°	22.9 mm

Image Size of Sun (mm) = Focal Length (mm) / 109

TABLE 22
SOLAR ECLIPSE EXPOSURE GUIDE

ISO	f/Number								
25	1.4	2	2.8	4	5.6	8	11	16	22
50	2	2.8	4	5.6	8	11	16	22	32
100	2.8	4	5.6	8	11	16	22	32	44
200	4	5.6	8	11	16	22	32	44	64
400	5.6	8	11	16	22	32	44	64	88
800	8	11	16	22	32	44	64	88	128
1600	11	16	22	32	44	64	88	128	176

Subject	Q	Shutter Speed								
Solar Eclipse										
Partial[1] - 4.0 ND	11	—	—	—	1/4000	1/2000	1/1000	1/500	1/250	1/125
Partial[1] - 5.0 ND	8	1/4000	1/2000	1/1000	1/500	1/250	1/125	1/60	1/30	1/15
Baily's Beads[2]	11	—	—	—	1/4000	1/2000	1/1000	1/500	1/250	1/125
Chromosphere	10	—	—	1/4000	1/2000	1/1000	1/500	1/250	1/125	1/60
Prominences	9	—	1/4000	1/2000	1/1000	1/500	1/250	1/125	1/60	1/30
Corona - 0.1 Rs	7	1/2000	1/1000	1/500	1/250	1/125	1/60	1/30	1/15	1/8
Corona - 0.2 Rs[3]	5	1/500	1/250	1/125	1/60	1/30	1/15	1/8	1/4	1/2
Corona - 0.5 Rs	3	1/125	1/60	1/30	1/15	1/8	1/4	1/2	1 sec	2 sec
Corona - 1.0 Rs	1	1/30	1/15	1/8	1/4	1/2	1 sec	2 sec	4 sec	8 sec
Corona - 2.0 Rs	0	1/15	1/8	1/4	1/2	1 sec	2 sec	4 sec	8 sec	15 sec
Corona - 4.0 Rs	-1	1/8	1/4	1/2	1 sec	2 sec	4 sec	8 sec	15 sec	30 sec
Corona - 8.0 Rs	-3	1/2	1 sec	2 sec	4 sec	8 sec	15 sec	30 sec	1 min	2 min

Exposure Formula: $t = f^2 / (I \times 2^Q)$ where: t = exposure time (sec)
f = f/number or focal ratio
I = ISO film speed
Q = brightness exponent

Abbreviations: ND = Neutral Density Filter.
Rs = Solar Radii.

Notes: [1] Exposures for partial phases are also good for annular eclipses.
[2] Baily's Beads are extremely bright and change rapidly.
[3] This exposure also recommended for the 'Diamond Ring' effect.

F. Espenak - 2001 Aug

REQUEST FORM FOR NASA ECLIPSE BULLETINS

NASA eclipse bulletins contain detailed predictions, maps and meteorology for future central solar eclipses of interest. Published as part of NASA's Technical Publication (TP) series, the bulletins are prepared in cooperation with the Working Group on Eclipses of the International Astronomical Union and are provided as a public service to both the professional and lay communities, including educators and the media In order to allow a reasonable lead time for planning purposes, subsequent bulletins will be published 18 to 24 months before each event. Comments, suggestions and corrections are solicited to improve the content and layout in subsequent editions of this publication series.

Single copies of the bulletins are available at no cost and may be ordered by sending a 9 x 12 inch SASE (self addressed stamped envelope) with sufficient postage for each bulletin (12 oz. or 340 g). Use stamps only since cash or checks cannot be accepted. Requests within the U. S. may use the Postal Service's Priority Mail for $3.95. Please print either the eclipse date (year & month) or NASA publication number in the lower left corner of the SASE and return with this completed form to either of the authors. Requests from outside the U. S. and Canada may use ten international postal coupons to cover postage. Exceptions to the postage requirements will be made for international requests where political or economic restraints prevent the transfer of funds to other countries. Professional researchers and scientists are exempt from the SASE requirements provided the request comes on their official or institutional stationary.

Permission is freely granted to reproduce any portion of this NASA Reference Publication All uses and/or publication of this material should be accompanied by an appropriate acknowledgment of the source.

Request for:	NASA TP-2001-209990 — Total Solar Eclipse of 2002 December 04

Name of Organization:	_____
(in English, if necessary):	_____
Name of Contact Person:	_____
Address:	_____
	_____
City/State/ZIP:	_____
Country:	_____
E-mail:	_____

Type of organization:	___ University/College	___ Observatory	___ Library
(check all that apply)	___ Planetarium	___ Publication	___ Media
	___ Professional	___ Amateur	___ Individual

Size of Organization: _____ (Number of Members)

Activities:	_____
	_____
	_____

* *

Return Requests	Fred Espenak	or	Jay Anderson
and Comments to:	NASA/GSFC		Environment Canada
	Code 693		123 Main Street, Suite 150
	Greenbelt, MD 20771		Winnipeg, MB,
	USA		CANADA R3C 4W2

	E-mail: espenak@gsfc.nasa.gov		E-mail: jander@cc.umanitoba.ca
	Fax: (301) 286-0212		Fax: (204) 983-0109

2001 Sep

www.ingramcontent.com/pod-product-compliance
Lightning Source LLC
Chambersburg PA
CBHW081734170526
45167CB00009B/3817